Greenhouse

&

Hydroponics Gardening

2 Books in 1

A Step-by-Step Guide for Beginners on All You Need to Know to Grow Healthy Vegetables, Fruits & Herbs All-Year-Round with the Best Gardening Systems

Greenhouse Gardening

A Step-by-Step Guide for Beginners on Everything You Need to Know to Build a Perfect and Inexpensive Greenhouse to Grow Healthy Vegetables, Fruits & Herbs All-Year-Round

Table of Contents

Air Circulation

Cooling

Controllers/Automation

High-Pressure Sodium Fixtures

Fixed and Programmable Spectrum Led Fixtures

Ceramic Metal Halide

T5 Fixtures

What Is Irrigation?

What Is Drip Irrigation?

The Greenhouse Irrigation Strategy

Seasonal Temperatures

Seasonal Light Levels

Crop Development

Dry-down Levels

Running Water: Where Water Travels in A Greenhouse

Holding Tanks

Mixing in Nutrients

Water Distribution & Collection

Benefits of Greenhouse Irrigation System

Heating

Conduction

Convection

Radiation

Factors Affecting Heat Loss

Heat Loss Measurements

Pest Control Equipment

Greenhouse Equipment and Accessories Soil Sterilizers

Gardening Sieve – Sowing Sieve

Plant Support Equipment

Introduction

Congratulations on purchasing *Greenhouse Gardening: A Step-by-Step Guide for Beginners on Everything You Need to Know to Build a Perfect and Inexpensive Greenhouse to Grow Healthy Vegetables, Fruits & Herbs All-Year-Round*, and thank you for doing so.

I am glad that you have chosen to take this opportunity to discover the full potential and benefits of greenhouse gardening.

I am sure that the information you will find in this book will guide you step by step, and you will learn in a fun and straightforward way how to approach gardening.

Greenhouse gardening is a rewarding way to grow your vegetables, herbs, and fruit.

The best thing about a greenhouse is that you don't need a piece of land to grow amazing crops.

Greenhouses use the sun to create a warm and beneficial environment for growth; they also protect plants from the weather, pests, diseases, and prolong growing seasons. We'll see that there are many types of greenhouse structures, and you'll find the type that best suits you and your needs.

Thanks to the information you will find in this book, you will learn how to grow plants, fruits, and vegetables all year round in a controlled environment, without having to worry about the cold winter months.

Chapter 1: Introduction to Greenhouse Gardening

The greenhouse is simply a building where plants are grown. Such buildings can either be small structures, or they can also be enormous in scale. The idea behind the greenhouse's dates back to Roman times when Emperor Tiberius demanded to eat Armenian cucumber every day, for which his gardeners had to use a device similar to those of modern greenhouses to make sure they had one every day.

Italy was the place of the first modern greenhouses in the 13th century. Initially, the greenhouses were more popular on the grounds of the rich, but soon they also branched out to universities. Some of the largest greenhouses ever constructed in the 19th century, while the 20th century pioneered the geodesic dome for use in several greenhouses.

How Does It Work?

The greenhouse effect, as it applies to the real greenhouses, functions as follows. The greenhouse decreases the rate at which thermal energy flows out of its system, and does so by impeding the heat consumed by convection from escaping its confines. The material used for building a greenhouse is usually glass or plastic so that sunlight can move through it. This sunlight is an integral part of the greenhouse being dry because it heats the ground within the greenhouse. On the other hand, the warm ground then warms up the air in the greenhouse, which prevents the plants from heated inside as it is limited to the structure of the greenhouse.

Uses of Greenhouse

The purpose of the greenhouse is to protect crops from excess cold or heat and unwanted pests. A greenhouse makes it possible to grow other types of crops during the year, and bananas, tobacco plants, vegetables and flowers are the most commonly grown greenhouses. High-level countries are where greenhouses are most common; this has to do with concerns about sustaining a viable food supply. For example, Almeria, Spain, is one of the world's largest greenhouses, spread over 50,000 acres.

Greenhouse Garden Gardening is one of the most common hobbies in the world, so running a greenhouse garden is just a logical extension. A greenhouse garden is mainly intended to prolong the growing season of valuable crops and plants. Horticulture enthusiasts will also be excited about the greenhouses, as it helps them to grow plants and flowers throughout the season, which can then be brought into the home. A greenhouse garden can be designed cheaply or inexpensively, with plastic or glass, and look elegant or merely practical. After selecting an excellent location for a greenhouse garden, you can create one yourself by ordering a greenhouse kit from any number of famous producers. These kits are self-made projects and can be as complicated or plain, or as large or small as you like.

Popular Greenhouses

Some of the most prominent greenhouses have exemplified what the operation of a greenhouse is all about, or are attached to research facilities that are doing productive work in the field of botany. One popular greenhouse is the Kew Gardens in England, which consists of 121 hectares of greenhouses and gardens. This facility is both studying and receiving visitors. Another prominent greenhouse is also based in England: the Eden Project. It is a community of geodesic dome greenhouses whose purpose is to educate people about the dependence of humanity on plants. Another famous greenhouse named Glass City is located in the Netherlands; it is a large number of greenhouses situated in the largest greenhouse area in the Netherlands, called the Westland.

Chapter 2: Different Types of Greenhouse

It would indeed be a very demanding gardener who would not be able to find a greenhouse entirely like the vast ranges that are now available. There is almost an enormous abundance of choices between the cheapest and the most expensive ones.

Before making this vital decision (even the smallest, most modestly designed greenhouse would mean quite a bit of expense), it is required to discuss the different types of greenhouses in the market and the purposes for which they are most suitable.

Types of Greenhouse

A greenhouse should be placed where it is safe from frost and wind, but away from trees that may provide too much shade and drop leaves and branches on the structure. Find a spot that lets you have full sunshine. The south or southeast part of the building is usually the most desirable location. If this is not feasible, the morning sunlight on the east side may be adequate. Greenhouses with a ridge on top should be built in such a way that the ridge runs from east to west. The site also needs good drainage, even if that means raising the floor so that rainwater and irrigation will drain away. Other considerations include the location of water, heat and electricity. Greenhouses may be purchased in non-assembled kits; prefabricated and ready-to-assembled; custom-designed and constructed by professionals, or constructed by homeowners. Generally, greenhouse forms fall into three groups, with variations.

1. Span-Roof Greenhouse

The even-sided, free-standing, span-roofed greenhouse is perhaps the most popular type, making it possible to grow as many plants as possible under the best feasible conditions.

There are various variations on the theme, but the span-roofed glass-to-ground-level greenhouses are generally speaking. This type of greenhouse is best suited to average gardeners. These structures are suitable for edible crops, such as lettuce and tomatoes, and for decorative plants, such as chrysanthemums and carnations, which can be grown in beds on the floor of the house. Span-roofed greenhouses of this kind are equally satisfying for growing pot-grown plants on an ash or gravel foundation. The most significant advantage of this type of house is that the plants freeze the full amount of light. It must be said, however, that plants are less quickly taken care of when they are grown on the floor rather than on staging. Furthermore, the heat loss from a full glass greenhouse is greater than when the sides of the house are made of a low brick or wooden wall.

If the aim is to grow plants on staging, then the span-roof house with low walls, as I have just mentioned, is more fitting. This type is a typical span-roof house with a door at one side and staging on each side at the height where the walls end and the windows begin. A useful agreement between the two types already described is the design that allows the glass to be ground on one side, allowing the plants to be grown in beds, and the low wall and the staging on the other side, allowing the plants to be grown in pots.

2. The Dutch-Light Greenhouse

This type of greenhouse is cheaper to buy than others, as it consists basically of regular Dutch-light panels which are mounted together to make a greenhouse. Such a structure is noteworthy for the amount of light that can enter plants through large areas of unobstructed glass. Dutch-light greenhouses are suitable for gardeners whose primary interest is to grow tomatoes during the summer, chrysanthemums for late autumn and early winter and lettuce for early spring. This type of house is widely used by commercial growers and, in this case, what is good for them is also good for amateur gardeners-but, of course, on a much smaller scale. Today's Dutch-light house was much more draught-proof than its predecessors a few years ago. The lights match tightly together to avoid cold winds, and the roofs are built in such a way as to prevent flooding, which, as any gardener knows, can be as much resented by plants as by draughts.

Another benefit of this type of house is its versatility, as extra lights can be attached to the standard unit to make up a house that is exactly in line with one's requirements (Perhaps 1 can also suggest that some versatility can be built into certain span-roof designs, with extensions available for standard units.) Dutch-lights are made of softwood or Canadian Red Cedar. I would suggest purchasing the latter, although they are slightly more expensive since they last a lot longer and need very little care

.

3. The Lean-To Greenhouse

The lean-to-type greenhouse uses the existing wall, thereby shaping one side of the frame and reducing the cost of materials. Because the wall is usually a house wall, there is likely to be some heat transfer in the winter, which is another benefit. In certain situations, the wall will hold the heat much longer than the windows, and the greenhouse itself will be less open to the elements than a free-standing building.

The lean-to type of greenhouse is suitable for a south-westerly or south-easterly facing wall, and the wall itself can be used to an excellent opportunity for growing a trained peach or nectarine tree or any of the many climbers or flowering shrubs that make such an attractive appearance when trained against a wall.

Like a span-roofed greenhouse, the lean-to-type may have glass to the ground or have a low wall or an embroidered foundation to the height of the staging. This type of greenhouse, a combination of span-roof and lean-to-shaped designs, is not commonly seen today, although it has many advantages. Greenhouses of this sort are designed against a wall like a lean-to-greenhouse, except in this case the house is higher than the building.

4. The Circular Greenhouse

A beginner in the gardening scene is the circular greenhouse. This offers more space for plants in a given area than other types of greenhouses. In smaller gardens, where space is essential, this could be an especially vital factor. The conservatory or The Sun Lounge, the Modem Sun Lounge (updated conservatory), ideally with access to the living room, has a benefit that is becoming more and more recognized. Many plants which need only protection against frost can be grown in such a structure, and the sun heat will provide appropriate sitting-out conditions for many days in the winter, particularly if there is a steady transfer of heat from the house through an open door or a French window. It is very normal that (the sun lounge remains unsealed except when the weather is coldest, while a small heater can be used to hold the temperature above the freezing mark.

Greenhouse Door Hinged or Sliding?

A lot of the modem greenhouses have sliding doors rather than hinged doors. These may be of great use if space is at a premium. A slight problem with sliding doors, however, is that the base runners can be hampered by grit. Both doors should be wide enough to allow easy access to the wheelbarrow because there are many occasions when this equipment element is required to transport items like manure, plants and flower pots.

Chapter 3: Planning and Building Your Greenhouse

Getting Started

As with any significant purchase, several vital decisions must be made at the beginning of the purchase process. The first choice, of course, is the amount that you are willing and able to invest in your greenhouse. Next, you should know how you're going to use it. Is it going to be over winter plants, start crops, or cultivate orchids all year round? Next, you will decide which greenhouse is best suited to your needs. Do you need a portable or more permanent structure? Do you want a frame that doesn't require maintenance, or are you prepared to take extra care of the wood structure? Is tempered glass or polycarbonate the best cover for you? What's the right side of the greenhouse for you? In the end, looks should be remembered. Does your district or licensing agency have special requirements? Once you've been through this cycle of learning, you're good to go ahead.

Location

The greenhouse should be placed where maximum sunlight is received. The first option of location is the south or southeast side of a building or shade tree. Sunlight is best all day, but morning sunlight is most beneficial because it helps the plant's food production cycle to begin early; thus, growth is maximized. The east side position catches much of the sunshine from November to February. The next best positions are southwest and west of the main buildings, where the plants will obtain sunshine later in the day. North of the main structures is the least suitable position and is ideal only for plants needing limited light.

Deciduous trees, such as oak and maple, can beneficially shade the greenhouse from the severe afternoon summer sun; nevertheless, they should not shade the greenhouse in the morning. Deciduous trees often require full exposure to the winter sun, as they shed their leaves in the fall. Evergreen trees that have foliage all year round should not be placed where they can shade the greenhouse because it will block the less powerful winter light. You should strive to maximize the exposure of the winter sun, particularly if the greenhouse is used all year round. Note that in the winter, the sun is lower in the southern sky, allowing the buildings and evergreens to cast long shadows.

Good drainage is also a requirement of the site. When required, create a greenhouse above the surrounding ground so that rainwater and irrigation can drain away. Other site considerations include light requirements for plants to be grown; location of sources of heat, water and electricity; and protection from the winter wind. Connection to the greenhouse should be convenient for both humans and utilities. A potting plant and a storage area for supplies should be located nearby.

Foundations and Floors

Few of the greenhouses either come with or have a foundation available to build on. The others can be constructed on treated lumber frames, but we prefer a more durable foundation such as poured concrete, bricks or blocks similar to those in residential buildings. The foundation applies only to the perimeter of the greenhouse.

Permanent flooring is not recommended because the soil mix media may remain wet and slippery. A concrete, gravel or stone walkway, 24 to 36 inches wide, can be built for easy access to plants. The remainder of the floor should be filled with a few inches of gravel to remove surplus water. Water can as well be sprayed on the gravel to create moisture in the greenhouse. If a permanent floor is required, good drainage and anti-slip surface should be given.

Glazing

These are only representative data for light transmission and U values. The exact values of the sheets will depend on the manufacturers of the glazing materials. Polycarbonate usually has a 10-year warranty against yellowing and brittle, and hail damage. We had sheets looking good for 15 years, and we had news that 20-year-old polycarbonate still works. Polycarbonate is a multiwall material that gives you additional insulation over a single glass. The only downside that we consider to be polycarbonate is the consistency of the material. The rib that runs through the middle of the sheets will distort your vision. You will see the color of the leaves and the flowers, but not visible enough to differentiate which sort of flower you see.

Product Light Transmission / U- Value

Double clear glass 82 /.55

Double clear glass W/ low E 75 /.35

16mm Polycarbonate Clear 78 /0.545

8mm Polycarbonate Clear 80 /0.635

6mm Polycarbonate Clear 80 / 0.645

Single clear glass 90 / 1.11

Heating

The heating requirements of the greenhouse depend on the ideal temperature of the plants grown, the location and design of the greenhouse and the total outside area of the structure. As much as 25% of the daily heat requirement can come from the sun, but a poorly insulated greenhouse structure may need a lot of heat on a cold winter night. The heating system must be sufficient to maintain the appropriate day or night temperature.

Solar heater greenhouses have been popular for a short time during the energy crisis, but have not proved to be economical to use. Different solar collection and storage systems are large and need a lot of space. However, greenhouse owners should experiment with heat-collecting methods to reduce fossil fuel consumption. One method is to paint black containers to attract heat and to fill them with water to keep them in place. However, since the temperature of the greenhouse air must be maintained at rising plant temperatures, the greenhouse itself is not a great solar heat collector.

Calculating the Heating System

Heating systems are measured in British Thermal Units (Btu) per hour (h). The Btu of the heating system, Q, can be calculated using three factors:

1. A is the total area of the exposed greenhouse sides, the ends, and the roof is one unit; on the curved eave, the sides and the roof are one unit; measure the length of the curved support beam and multiply by the length of the building, multiply by two on the free-standing curve. The curve end area is 2 (ends) X 2/3 X width. Add the sum of the first measurement to the second.

2. U is a heat loss factor that quantifies the rate at which heat energy flows from the greenhouse. For example, a single glass cover has a value of 1.2 Btu / h x (ft x ft) x degree F (heat loss in Btu per hour per square foot area per degree in Fahrenheit). The values allow some air infiltration but are based on the assumption that the greenhouse is relatively airtight.

3. (Ti – To) is the average difference between the lowest outside temperature (To) in your region and the temperature to be maintained in the greenhouse. For example, the greatest difference would occur early in the morning with an outdoor temperature of 0 to -5 degrees Fahrenheit. In contrast, an indoor temperature of 60 degrees is maintained. Plan a temperature difference of between 60 and 65 degrees. The following equation summarizes this definition Q = A x U x (Ti – To) This explanation is a little technical. Still, these considerations must be considered when selecting a greenhouse. Note the influence of each value on the outcome. When different materials are used in wall or roof construction, the heat loss for each material must be calculated. For electrical heating, transform Btu / h into kilowatts by dividing Btu / h by 3,413. The fresh air inlet is advised when using wood, gas or oil in the greenhouse. Unvented heaters are not approved for use.

Air Circulation

The installation of revolving fans in your greenhouse is a safe investment. During winter, when the greenhouse is heated, you have to keep the air in circulation so that temperatures stay constant in the greenhouse. Without air mixing fans, the warm air rises to the top, and the cool air falls around the plants on the field.

Mini fans with a cubic foot per minute of air moving capacity equal to one-quarter of the greenhouse air volume are adequate. For small greenhouses with a length of less than 60 feet, position the fans diagonally opposite corners, but out from the sides and the ends. The goal is to establish a circular (oval) pattern of air movement. Switch the fans off during the summer when the greenhouse is required for ventilation.

The ventilator in a forced-air heating system may also be used to provide continuous air circulation. The fan must be connected to the on / off switch so that it can run continuously, separately from the thermostatically controlled burner.

Cooling

Ventilation is essential even in cold weather conditions. The greenhouse can get too warm for bright, sunny days. Ventilation is an exchange of indoor air for outdoor air to regulate temperature, extract moisture, or replenish carbon dioxide. Many ventilation systems can be used. Be cautious when combining sections of two structures.

Natural ventilation uses ridgeline roof vents with side inlet vents (louvers). Warm air rises on convection currents to exit through the rim, pulling cool air from all sides.

Mechanical ventilation uses an exhaust fan to convey air out of one end of the greenhouse while outside air goes into the other end through motorized inlet lovers. Exhaust fans should be sized so that the total volume of air in the greenhouse can be changed every minute.

The total amount of air in a medium to large greenhouse can be determined by multiplying the floor area by 8 times (the average height of the greenhouse). A small greenhouse (less than 5000 cubic feet in air volume) would have an average exhaust fan capacity by multiplying the floor area by 12.

The exhaust fan capability should be chosen at one-eighth of an inch of static water pressure. The static pressure rating is for air resistance by lovers, fans, and greenhouses and is usually displayed in the fan selection table.

The criteria for ventilation vary with the environment and the season. One must determine how much of the greenhouse is going to be used. In the summer, 1 to 1.5 changes in air volume per minute are expected. Small greenhouses need more. In winter, 20-30% of one volume of air exchange per minute is adequate to mix in cold air without cooling the plants.

This condition cannot be met by a single-speed fan. Two-speed fans are better off. The combination of a single-speed fan and a two-speed fan allows three ventilation speeds that better suit the needs of the year-round. To control the operation, a single-stage and a two-stage thermostat are required.

A low-speed two-speed engine provides around 70% of its full capacity. When two fans have the same capacity rating, the low-speed fan delivers around 35 per cent of the total. This rate of ventilation is reasonable for the winter. The fan runs at high speed in the morning. Both fans are operating at high speed during the summer.

Some greenhouses are sold by a manual vent or ridge vent. The manual machine can be a backup system, but it doesn't take the place of a motorized one. Do not take short cuts when designing an automated control system.

Shade Cloth is one of the essential elements to keep the greenhouse cool in the summer. It can be mounted inside suspended from wires or outside either fastened to grommets on the edge of cloth with ropes or using a roll-up and down system. It comes with a 30-90 per cent light transmission. The most efficient heat reduction when built outside, but here it interferes with the vents. Some blinds can be mounted with the same issue of interrupting the ventilation as with the shade cloth on the outside. The inside blinds can be adjustable. The temperature of the ventilation system inside the greenhouse should be between 3 and 4 degrees below the outside temperature.

Misting systems: There are several systems available that can be used for cooling and raising relative humidity in the greenhouse.

Relative humidity (RH) is defined as the measure of how much water is dissolved in the air at a specific temperature expressed as a percentage. In general, the growth of many plants is relatively unaffected by HR between 45 and 85 per cent. Plants growing at RH below 45 per cent may grow slowly, have smaller leaves, require more water, or develop burnt leaf margins or leaf tips. Plants growing at RH above 85 per cent are susceptible to fungal pathogens, particularly if the foliage is condensed with water.

Several conditions can exist in a greenhouse that causes problems due to high or low HR. High light, high temperature, and rapid ventilation of the air can reduce RH to unacceptable levels during the summer. Shading to reduce light and temperature and the use of mysteries or evaporative cooling is the best solutions. It is a good idea to keep the greenhouse full of plants because the plants produce a lot of RH.

Evaporative cooling systems are available in two versions, a small kit of evaporative coolers and a fan and board. A small package of coolers has a fan and pad in one box to evaporate water, which cools the air and reduces the humidity. Heat is removed from the air to transform water from liquid to vapor. Moist, cooler air enters the greenhouse while the heated air moves through roof vents or exhaust lovers. The other system is used in commercial greenhouses, places the pads on the air inlets at one side of the greenhouse and uses exhaust fans at the other end of the greenhouse to pull air through the room. The evaporative cooler works more effectively when the humidity of the outer air is low. On a hot sunny day, the humidity level of the atmosphere remains nearly constant. This means that the RH is lower in the afternoon when the temperature is at its highest. And the lower the moisture, the better the evaporative cooling effect.

Simply put, the cooling effect is best if you need it most. Size the cooler evaporative capacity at 1.0 to 1.5 times the volume of the greenhouse. The temperature inside the house can be as much as

10 to 15 degrees cooler than the temperature outside with a properly designed device.

Controllers/Automation

Automatic control is necessary for the maintenance of a healthy greenhouse environment. On a winter day with unstable amounts of sunlight and clouds, temperatures could fluctuate greatly; close monitoring would be needed if a manual ventilation system were in use. Therefore, both hobbyists and commercial operators should have automated systems with thermostats and other sensors, unless close monitoring is possible. An alarm system is also very critical if the system fails.

Thermostats can be used to control individual devices, or a single temperature sensor central controller can be used. In any case, the sensor or sensors should be shaded from the light, placed around the height of the plant away from the sidewalls, and have constant airflow over it. An aspirated box is recommended; the box contains each sensor and has a small fan that pushes the greenhouse air through the box and over the sensor(s). The box should be painted to reflect solar heat and allow accurate readings of the temperature of the air.

Planning and constructing a greenhouse is a delightful process. Only make sure you do your homework.

Chapter 4: The Pros and Cons of Greenhouse Farming

Before diving into greenhouse farming, like any other type of farming, it is essential to consider the pros and cons to be able to build a viable business plan around the practice that has been going on for centuries and is now in the hands of more people thanks to the ever-increasing technology in the agriculture industry. These are some advantages and disadvantages of greenhouse farming.

The Pros of Greenhouse Farming

1. Increased Production

Greenhouse farming is considered to be the implementation of intensive agriculture and may contribute to an increase in crop production. This is so because you have more control over establishing the optimum climate conditions required for plant growth and can produce more plants per square foot compared to open field crops.

2. Minimizing Production Risks

In the enclosed environment, it can help prevent crops from being affected by climate change-related incidents, such as sudden rises or drops in temperature, as well as keeping crops away from birds and other animals.

3. Maximizing Profits

Several studies have shown that profits per crop per square foot may be either twice or three times as high when introducing greenhouse farming as an alternative to open field farming and when combining practice with other techniques such as Hydroponics. By using resources more efficiently, you can generate less waste, which in turn can result in higher profits.

4. Increased Pests, Weeds and Disease Control

A properly designed and optimally built greenhouse can prevent problems such as pests and weeds and provide more control over other diseases. The enclosed space can only be restricted to the necessary personnel, and fewer people entering and leaving means less risk of bringing unwanted elements closer to the crops.

5. Ability to Grow Year-Round Produce, Even Off-Season

A greenhouse is entirely separate from the outside world, removing the constraint of growing crops for a particular season only. High-quality crops can be grown even in harsh winters or high summer temperatures, given that you have the means to build the right environment inside the greenhouse.

6. More Stability and Security

Because you do not rely on climate conditions, an increase in stability and protection, not only for crops but also for workers, can be achieved through greenhouse farming.

The Cons of Greenhouse Farming

1. You Need A Sizeable Initial Investment

The construction and design of the house are relatively costly, which can be difficult for many farmers. Greenhouse farming is recommended for profitable crops that are easy to market to maximize the chances of a quick return to your investment.

2. Specific Greenhouse Design

A greenhouse must be planned and designed to take into account exact details and elements such as the location, the type of crops that will be grown on it and the type of technical add-ons that will be needed. If the greenhouse is not correctly built from the start, this could affect the desired results and would also mean spending more money down the line.

3. High Production Costs

Generally speaking, the operating costs of greenhouse farming are higher than those of open-field agriculture. Maintaining the right conditions for plant growth within the greenhouse means spending money on electricity or gas, to give an example.

4. Higher Skill Level Required

Within the greenhouses, the employees are entirely responsible for the plants, being able to monitor all environmental factors means that any issues that occur are caused by people and should be resolved immediately. This includes qualified and skilled professionals who can ensure that the operation can be carried out efficiently and safely.

5. Ideal Conditions for Disease

Just as the conditions within the greenhouse will be optimal for the growth of crops, the same can be said for diseases that may affect them. Although greenhouse farming can indeed provide better safety measures against these problems, if they are not correctly implemented, there is still a risk of pests being introduced into the greenhouse and growing faster than usual, which can cause production losses.

6. You Need an Established Market Operation

You need an existing business network to make the most of greenhouse farming. It is imperative that you have your distribution and sales system already in place. Being able to sell higher quality products at higher prices is one of the most significant benefits of greenhouse farming; these crops are short-lived and need to be sold quickly to ensure quality. The more time the commodity spends in storage, the more value it loses. Getting a commercial partner and a competitive market will be essential and can be a challenge for many.

Chapter 5: The Management of The Greenhouse Environment for a Quality Control

Your crop production can be significantly accelerated through the use of a greenhouse control or automation system. With this type of environmental regulation, the greenhouse remains constant to provide optimal conditions that are most favorable for maximum yield.

The ability of a plant to grow and develop depends mainly on photosynthesis. In the presence of light, the plant combines carbon dioxide and water to produce sugars that are then used for growth and flower/fruit production.

The management of the greenhouse environment is aimed at maximizing the photosynthetic cycle of plants, the ability of plants to use light at optimum output.

Greenhouse Lighting Control

There's a lot more to the decoration of the greenhouse that meets the eye. Growers looking for suitable lighting for their greenhouse should consider the following three factors: the type of crop being cultivated, the time of year and the amount of sunlight available.

Greenhouses generally require six hours of direct or full-spectrum light per day. If this cannot be done naturally, additional lighting must be included. Additional lighting is the use of multiple, high-intensity artificial lights to encourage crop growth and yield. Hobbyists like to use them to maintain growth and extend the growing season, while commercial growers use them to boost yields and profits.

Growers have a wide variety of lighting choices to choose from, so it is necessary to understand the complexities of different lighting types. Again, this is made easier to handle with greenhouse environmental restrictions that can be prepared and tracked.

Humidity Control

As plants begin to increase their growth rate, you may want to will moisture gradually to promote transpiration, allowing more water to flow through the plant. As the plant consumes more water, the elongated cells fill and add nutrients to the growing parts of the plant.

Humidity should also be controlled carefully because if it gets too hot in the greenhouses, plant leaves have a much higher chance of being wet. Sadly, damp leaves are one of the easiest ways to allow a fungal infection or a mildew outbreak. Fungal diseases such as Botrytis pathogen or powdery mildew are natural causes of greenhouse disease. Controlling and monitoring the greenhouse ecosystem ensures greater quality management.

Ventilation & Fan Control

The use of vents is another simple way to help control temperature and humidity. Through the use of rack and pinion and ventilation control, if it begins to get too hot, the greenhouse vents can be opened at a fixed temperature.

We also measure relative humidity (the amount of water vapor present in the air expressed as a percentage of the amount required for saturation) which can also be decreased by opening the vents. Warm dry air, and not wet.

Horizontal airflow fans with greenhouse control systems can also be triggered. These increase air circulation and help to extract moisture from the air. To maintain a proper balance, it is essential to increase the greenhouse temperatures.

With proper greenhouse temperature and humidity sensors, all of this is controlled by our greenhouse automation machine. This will help you track and control the humidity and temperature levels more effectively. A grower-approved greenhouse environmental control program ensures that all rates are kept in check.

Carbon Dioxide or CO2 Control

Supplying a lot of carbon dioxide to your plants is essential for healthy plant production. Plants take carbon from the environment as a necessary part of photosynthesis. Through the stomata openings, carbon dioxide enters the plant through the diffusion cycle.

CO2 increases productivity by increasing plant growth and overall health. Some forms in which CO2 productivity is increased include earlier flowering, higher fruit yields and longer growth cycles.

Going into some of the more advanced calculations here, but for most greenhouse crops, net photosynthesis increases as CO2 levels rise from 340–1,000 ppm (parts per million).

Most crops show that, for any given level of photosynthetic active radiation (PAR), increasing the level of CO_2 to 1,000 ppm would increase photosynthesis by about 50 per cent above the level of ambient CO_2.

Stay within the quality control guidelines of our greenhouse climate control system to control the CO_2 ranges properly.

Air Temperature Control

The increase in air temperature will increase the rate of photosynthesis to a point. However, above 85 degrees, the plants go into photorespiration. Plants can begin to wither, which is not ideal for growing.

If you do not balance higher air temperatures with higher levels of carbon dioxide and light intensity, your plants will do more photorespiration than photosynthesis, which will have an immense impact on the health of your plants.

At some stage, enzymes will not perform their functions and will fall apart, and your plants will not develop a healthy metabolism. Balance with a greenhouse temperature control system is essential in all aspects.

Regular Irrigation and Fertilizers – Fertigation Control

We like a regular formula for irrigation and nutrient feed for greenhouse crops. In today's big commercial operations, this is where finishing automation cannot only help but also help you plan your other farms.

Water and fertilizers are used continuously for fertilization at precise quantities through the irrigation system. Supplying the nutrients required by crops helps to keep yields at their best.

Fertigation is particularly useful in the case of drip irrigation. With our automated finishing equipment, the water and nutrients are absorbed directly into the roots, improving the growth rate, the resilience and the quality of the crops.

The system is a more rational use of water and fertilizers. I think that respect for the environment and mitigating the impact on the environment is something that we can all resolve. We can also use a greenhouse water recycling system to ensure the health of your crops.

Full Greenhouse Automation

Automation will play a significant role in your greenhouse environment and make your indoor growing output manageable.

A control technology enables growers to program their growing environment as they like it. Any changes in rising conditions can be easily reversed by automation. Advanced IIOT accredited systems allow growers to track their growing indoor environment from their smartphones or laptops.

Chapter 6: Growing in an Ideal Greenhouse Climate

Although it may be hard to create the ideal greenhouse climate for growing plants, there are variables that growers can manage to optimize plant growth.

Irrespective of the type of crop being grown in the greenhouse, the environment that the grower is trying to achieve includes regulation of the same variables.

Greenhouse farmers are trying to control temperature, humidity, light levels, carbon dioxide and, in some cases, airflow and air circulation. Both variables have different set points, depending on the crop. These may also have different acceptable maximum and minimum ranges or rates.

Depending on the crop, these factors can be changed at different times of the day. For example, a tomato crop wants an average temperature of around 72oF per day. If the plants experience high temperatures at day time, if the temperature can be cooled down during the night, as long as the average temperature is 72oF, the tomato plants will be happy. In the case of lettuce, the grower may not be able to manipulate the day or night temperature to account for the excess of the maximum temperature that occurs during the day or night. This is one way these crops differ. Usually, when it comes to controlling the greenhouse environment, growers focus first on temperature.

The first line of defense against hot temperatures is not shading. Growers are trying to maximize as much light as possible in the greenhouse. Solar production is decreased as soon as the shade curtain is closed. The first line of defense for the cooling of the greenhouse is ventilation, either natural or mechanical.

If the temperature the growers need cannot be reached by ventilation, then some form of cooling is added. Typically cooling is done through evaporative cooling. This could be wet pads and a fan system, high-pressure fog or a low-pressure misting system in combination with mechanical and natural ventilation. If that doesn't work, then a shade curtain can be pulled. A shade curtain is usually only drawn for two to four hours during the day. It's pulled during the peak solar heat gain period. A shade curtain can cut the temperature by 2°F-4°F.

The Challenge of Reducing Humidity

Between controlling of greenhouse temperature and humidity, humidity is the most challenging variable, especially when used for dehumidification.

If the grower is trying to extract moisture from the greenhouse, there are a lot of problems. The primary method for extracting moisture from the greenhouse is through ventilation. But this means that the level of moisture or heat outside the greenhouse is higher than that within the greenhouse.

If the grower is seeking to increase the humidity or humidification of a greenhouse located in the southwestern U.S. where it is scorched, moisture can be added to the greenhouse using evaporative cooling. Another advantage of evaporative cooling is a reduction in the temperature that cools the temperature of the greenhouse. In a dry climate, evaporative cooling works very well to do both of these things.

Growers in the Midwest and Southeast will face more challenging conditions because they have high solar heat gains like growers in the Southwest, but they often have high humidity rates that require ventilation.

The climates in the Southeast and Midwest make it challenging to grow plants in a greenhouse due to humidity. The only defense line for growing plants in this kind of environment is ventilation. Growers want to share as much air as possible with the outside to minimize moisture and heat throughout the day. Typically, this is not enough. If the outside temperature is 90oF and the relative humidity is 90 percent, growers do not want these situations in their greenhouses.

If the temperature and humidity are increases, growers do not have the opportunity to use evaporative cooling because they cannot reduce the temperature enough. They can shade the greenhouses, but they only lower the temperature by 20F-40F from outside conditions. If it is 90oF and 90 percent humidity, pulling shade results in 86oF and 90 percent humidity, and this will not provide the vapor pressure deficit that the grower is trying to achieve.

Growers may consider closing their greenhouses to avoid bringing in hot, moist air, but this creates additional challenges.

Closing the greenhouse will cause the greenhouse to heat up from the sun, while the plants release moisture, resulting in the greenhouse only getting hotter. So far, I haven't seen anyone come up with a cost-effective way to minimize heat and moisture. Indeed, the grower could use a refrigerant-based cooling system similar to an air-conditioning system that would provide dehumidification. But the size and complexity of these devices are prohibitively expensive.

Maintaining the Proper Vapor Pressure Deficit

Temperature and humidity are closely related by the vapor pressure deficit (VPD).

As long as a grower can regulate the greenhouse temperature, this generally means that he can monitor the humidity level to the point that the vapor pressure deficit is where it should be. Even if VPD is not the destination that growers are aiming for, this is the target they are trying to achieve with or without temperature regulation.

VPD is the difference between the quantity of moisture in the air and how much moisture the air can retain when saturated.

There's an optimum point for the VPD. For leafy greens and culinary herbs preferring lower VPD, the approved VPD range is 0.65 to 0.9 kilopascal (KPA) with an optimum value of 0.85 KPA. Tomatoes, cucumbers and peppers begin to get dryer. The VPD range for tomatoes is between 0.9 and 1.2 KPA.

There is more surface area for leafy greens for moisture to escape the plants. Plants tend to be in a more open environment, so they don't release too much moisture too quickly.

Providing Adequate Airflow

Airflow in the greenhouse is essential for breaking up the layer of moisture around the surface of the leaf.

If the leaves are water transpiring, the surface of the leaf itself is considered to be saturated. The surface of the leaf exchanges moisture with the air encompassing it. The more moisture in the air around the leaf surface, the lower the tendency to transfer moisture from the surface of the leaf to the air around it.

That's essentially what the vapor pressure deficit is all about. It is the contrast between how much moisture there is on the surface of the leaf at a given temperature and how much moisture there is in the air at the same temperature. If it is within the right range, the plants are satisfied as to the leaves freely share moisture with the air. If the vapor pressure deficit is very low, this means that the air has a lot of moisture in it, meaning that there would be less moisture transfer from the leaves to the air. Plants cannot transpire as quickly, and nutrients cannot be distributed to the rest of the plant as quickly as possible. If the vapor pressure deficit is very high, the air is dry, and the plants shut down. As a protection strategy, the plants will close down their stomata so that they do not breathe moisture into the air because it would happen too quickly. Loss of water by transpiration would have happened faster than the plants could have taken up water.

Horizontal airflow fans are the conventional way of generating ventilation and ventilation in a greenhouse.

Horizontal airflow fans usually are suspended from the trusses or the frame of the greenhouse and blow the air in a circular pattern over the tops of the plants without necessarily blowing directly down the plants. Just movement and motion are enough to generate turbulence to allow air mixing around the plants to induce transpiration and convection.

By breaking up the small saturation pocket of air around the leaves, it facilitates the transfer of moisture from the leaves to the air. In more humid conditions, when the air is circulated over the surface of the leaf, the grower may promote more transpiration from the plants than if there is no airflow. Airflow is one of those factors that is not discussed as often as temperature and humidity regulation. It's neglected.

With increasing interest in vertical farms, growers use large growing racks to try to create three-dimensional temperature and humidity conditions.

Under these states, it is very easy for the air to get trapped in the centre of the rack. Vertical farmers are conscious of airflow because they see these hot spots or wet spots in the middle of the rising rack, and they understand they need airflow.

It is the same circumstances as if plants are grown in a greenhouse. If more airflow is provided in the greenhouse, more moisture could be removed from the surface of the plant and help the plant cool by convection.

Maintaining the Proper Carbon Dioxide Level

Carbon dioxide is not significantly impacted by the outdoor climate, greenhouse growers are controlling it relative to the outdoors.

In a greenhouse where growers burn fuel to generate carbon dioxide and ventilate at the same time, there is a challenge as to how much carbon dioxide should be supplied and how should it be maintained? Is there a way to mitigate the immediate loss of carbon dioxide to the outside air through greenhouse ventilation?

One technique not to overuse carbon dioxide is to provide an enrichment boost to plants with carbon dioxide. Carbon dioxide can be supplied first thing in the morning during the first light before the opening of the greenhouse vents. Mainly, the plants take a deep breath as the sun starts to come out, and the stomata open. The sunlight or the extra lights are turned on, and the plants eat up the carbon dioxide. If the grower begins to ventilate as the humidity has increased overnight or the temperature continues to rise as the sun is rising, the carbon dioxide accumulation should be prevented so that it is not blown out of the greenhouse by the vents and drained by the fans. Some farmers use carbon dioxide enrichment all day as long as there is adequate sunlight or artificial light.

Growers can reduce the carbon dioxide loss by trying to deliver it as close to the leaves as possible.

Some growers use sub-floor or sub-bench ducts to deliver carbon dioxide. Some growers may use PVC tubing or fish tubing to disperse carbon dioxide through the crop and directly to the leaves. This is best if a grower can find a way to produce carbon dioxide effectively without getting in the way of all other equipment and people working in the greenhouse.

That's why some people are looking at the potential advantage of growing in vertical farms. There is enclosed space and, in most cases, it is done in buildings that are not leaky. Some growers have considered the greenhouses closed. The cannabis industry is interested in this, but the problem is that there is an outrageous energy bill to try and close the greenhouse and not use any ventilation or mechanical cooling.

Relationship Between Greenhouse Climate Variables

The optimum level of carbon dioxide for each crop varies. 700-1,500 parts per million carbon dioxides is the level that most farmers are seeking to use.

Carbon dioxide is most beneficial to plants when there is a lot of light and good temperature and humidity or a strong VPD. Carbon dioxide is transferred through the stomata of the leaf, the same as moisture through transpiration. At the right side of the VPD, the stomata are open to the limit and release moisture and gulp carbon dioxide.

The first thing is to have the right VPD to maximize stomata opening. The second thing is photosynthesis, which is light-driven. If the air is filled with carbon dioxide, but the light level is very low, a lot of carbon dioxide will be lost. There must be enough light to allow a high enough rate of photosynthesis, or the plants cannot use carbon dioxide. All three of these factors works together. A good VPD is required to open the stomata, an adequate level of light is needed for photosynthesis, and carbon dioxide is needed to maximize the cycle of photosynthesis.

Chapter 7: Understanding Greenhouse Lighting

There's a lot more greenhouse lighting than the eye can see. Growers looking for suitable lighting for their greenhouse should consider the following three factors: the type of crop being cultivated, the time of year and the amount of sunlight available.

Greenhouses typically allow six hours of direct or full-spectrum light every day. If this cannot be achieved naturally, external lighting must be used. Additional lighting is the use of multiple, high-intensity artificial lights to encourage crop growth and yield. Hobbyists like to use them to maintain growth and extend the growing season, while commercial growers use them to boost yields and profits.

Photoperiod control lighting is equally essential as additional lighting. A photoperiod of light is the number of hours that a plant receives light within 24 hours. For example, if the sun rises at 6 a.m. and sets at 8 p.m., a photoperiod of 14 hours has lapsed. Photoperiod control lights are used to simulate long days, cause early flowering or encourage delayed flowering, depending on plant needs.

Growers have a wide range of lighting options to choose from, so it is essential to understand the complexities of different lighting types. Let's look at the uses and advantages of four different types of lighting.

High-Pressure Sodium Fixtures

High-pressure sodium fixtures provide more orange and redder spectrum light and give the human eye a golden-white appearance. As they promote budding and flowering, they are generally used later in the growth cycle of the plant. These fixtures are approximately seven times more effective than incandescent bulbs and work best when used in combination with natural daylight, making them an excellent choice for greenhouses. High-pressure sodium lights also offer the potential for a 10% increase in intensity and photo-synthetically active radiation (PAR). They give high-pressure sodium lights about 4 to 5 minutes to warm up and one minute to cool down. That's why they're not perfect at places where lights turn on and off regularly. It is also essential to be cautious of the placement; high-pressure sodium lights should be placed 30 to 36 inches above the plant for optimal performance.

Fixed and Programmable Spectrum Led Fixtures

LED (light-emitting diode) fixtures are the longest-lasting choice provided by Growers Supply, with a typical lifespan of 50,000 hours. The LED diode won't flame out as quickly as regular light bulbs, which gives it an extremely long-life cycle. LED light fixtures are more efficient than standard lighting because more of the power input goes to light than to heat. For example, incandescent bulbs are only about 20% effective, as most of their input power is used to generate heat.

Perhaps one of the biggest benefits of LED lighting is significant energy savings. They are quickly incorporated into any project and deliver up to 70% savings compared to high-intensity discharge (HID) lighting.

There is no time required to warm up with an LED fixture, and they are also free of mercury, making disposal much more comfortable than other bulbs. LEDs provide superior functionality when used as the sole source of lighting, making them an attractive option for many growers.

Ceramic Metal Halide

Ceramic metal halide lamps are utilized for their blue light, even though they appear bright white to the human eye. They can easily act as a primary light source, with an estimated lifetime of between 8,000 and 15,000 hours. Because metal halides are 3 to 5 times more effective than incandescent bulbs, they are an excellent choice for areas that do not receive natural sunlight.

It is important to remember that metal halides need to be warmed up for around 5 minutes or less before they can give out maximum light. They do need a cool downtime of around 5 to 10 minutes before restarting. For this reason, they are not suggested for locations where the lights are frequently turned on and off.

Ceramic metal halide lights should be placed 30 to 36 inches above the plants and may darken leaves and the overall good-looking greenery. A Growers Supply PAR Lucent Ceramic Metal Halide Lights are perfect for use in greenhouse and hydroponic applications. Growers also use them in the early stages of plant life while seeds are in the vegetative growth process. The dimmable ballast enables growers to obtain the perfect light for their operation. They're quiet, too, so there's no annoying humming, clicking or high-pitched noise to contend with.

T5 Fixtures

T5 fixtures are the most effective and common choice for through fluorescent greenhouse lighting. They use more limited energy than traditional lamps and can last up to 50,000 hours. These environmentally friendly lights also feature aluminum reflectors for optimum efficiency. They are best suited for use in hydroponics, greenhouses, farms, barns and more. They can be used from the beginning of the seed phase to full-term development.

The letter "T" indicates the tubular shape of the lamp, and the number 5 denotes the diameter of the lamp in eighths of an inch. T5 lamps are thin, only 5/8 "inch in diameter, making T5 fluorescent tubes more effective than conventional fluorescent tubes.

GrowSpan's High-Performance 45 "T5 Fluorescent Lamp highlights exceptionally high lumen output and full-spectrum lighting that is excellent for plants from the seedling stage to full-term growth. Its minimal heat output suggests that it can be placed very close to plants, within 6 to 12 inches, to be precise. While technically there is nothing as too much light, it is essential not to use too much light in a small space, which can cause the surface area of the leaf to overheat.

With so many greenhouse lighting choices customized to different plant types and growth stages, it's easy to see why expert advice is so respected by growers everywhere. When working with GrowSpan, growers have access to custom lighting plans, automated and computer-controlled lighting and even Greenhouse Specialists providing professional lighting design. GrowSpan's wide range of lighting solutions, from additional growth to photoperiod control make the optimization of rising space an easy, efficient process.

Chapter 8: Greenhouse Irrigation System

Greenhouse-grown plants (and all plants, for that matter) cannot survive without some form of irrigation. While different types of greenhouse irrigation systems are accessible – including sprinklers and subsurface systems – we use a drip irrigation system because of its productivity and accuracy. Take a closer look at how the drip irrigation system operates and what tells crop irrigation plan!

What Is Irrigation?

But first of all, what is irrigation and why is it important?

Irrigation is an agricultural method that applies water to crops at appropriate intervals and quantities. When plants are irrigated with water, one of the essential elements required for the photosynthesis process is given.

Photosynthesis is a mechanism in which plants grow their food. To finish the photosynthesis process, the plant needs a combination of different factors, including light, carbon dioxide (CO_2) and water. When these elements enter the plant, the light will activate reactions in the structures inside the plant cell called chloroplasts, resulting in the plant's glucose energy. At this time, oxygen is also going to be released into the atmosphere for us to breathe!

Water is an essential part of the process of photosynthesis. That is why it is so necessary for a greenhouse operation to develop a strong irrigation strategy that delivers the optimal amount of water to its crop at the right time and in the right volume.

What Is Drip Irrigation?

Drip irrigation is precisely what it sounds like – irrigation administered to a crop by dripper technology. Every plant has its dripper tube, which is typically placed in the back corner of the substrate block where our plants are located. It is also necessary to ensure that the dripper is not forced too deep into the block – making it easier for the roots of the plant to access the nutrient water.

Drip irrigation is an effective way to irrigate greenhouse crops – the location of the dripper tubes makes it easy for growers to deliver precise quantities of nutrient water directly to the root systems of the plants.

Drip irrigation process uses a pressure-compensated drip tube system that releases the specified volume of water once x water pressure is applied to the drip valve. This allows water and nutrients to be evenly distributed across the entire crop.

With greater efficiency and precision than other irrigation systems, setting up a drip irrigation strategy made the most sense to us. After all, as a commercial greenhouse operation, we have hundreds of thousands of plants to irrigate every single day!

The Greenhouse Irrigation Strategy

When deciding an irrigation strategy, Growers have several variables to consider, including seasonal temperatures, light levels, crop production and dry-down levels.

Seasonal Temperatures

When temperatures fluctuate at different times of the year, this has a strong impact on irrigation strategy. In summer months, when temperatures are very high and are more difficult to regulate internally, we usually need to irrigate the plants more frequently as they require higher amounts of water. On the other hand, we do not irrigate the plants as much during the winter months, because the internal temperatures are easier to maintain in the winter, and the plants usually do not need as much water.

Seasonal Light Levels

Just as temperature indicates irrigation needs, so do light levels – the more sunlight shining on plants, the more commonly they need to be irrigated. This implies that when the sun is out for a more extended period during the summer, the plants need more water, and when the days are shorter during the winter months, we don't need to irrigate the plants as often as possible.

Remember the process of photosynthesis – light is what activates the process of energy generation of the plant, so more light means that the plant will start working harder to grow leaves and fruit (which means that more elements like water and CO_2 are needed if the plant succeeds in generating glucose to feed itself).

Crop Development

The needs of the crop always change as it goes through different stages of ripening, which means that our irrigation strategy needs to shift with it! It is up to growers to track the production of their crop so that the required changes can be made to the irrigation strategy based on the ripening status of the crop.

Dry-down Levels

Dry-down refers to how much a substrate (which, in our case, is coconut fiber) has dried. Dry-down rates help inform our day-to-day irrigation strategy. If the coconut fiber block has dried out, it will weigh less and probably require irrigation – but if the coconut fiber slab is still full of moisture, it will weigh more and probably not be irrigated immediately. Growers use a growing scale that weights the plant base to measure dry-down rates.

Once a strategy has been developed, growers use PRIVA to change the irrigation settings. Growers can either upgrade the system to PRIVA terminals stationed throughout our facilities, or to their workstation computers or smartphones.

Running Water: Where Water Travels in A Greenhouse

From the moment we pump clean municipal water into our greenhouses, Irrigation water is starting an exciting journey!

Holding Tanks

Within each greenhouse, there should be three large water storage tanks, which contain clean water, polluted water and filtered water. When we pump clean municipal water to the greenhouses, the water is stored in a clean water tank.

Use a closed-loop irrigation system, which ensures that the excess water that our plants do not use in their root systems is collected, checked, washed and sent back to the plants. When you collect this excess water, it is stored in an untreated water storage tank until it is ready to be tested, cleaned and sent back to the plants. When treated, it is stored in the treated water storage tank until the time has come for irrigation.

Mixing in Nutrients

Ensure that the plants get the right amount of nutrients they need by mixing variety-specific, nutrient-rich fertilizers in large mixing tanks.

For optimal growth, each plant variety needs a specific amount of different nutrients. However, almost all fertilizer recipes include nutrients such as magnesium, iron, nitrogen, potassium and calcium.

When it is time to irrigate the plants, a mechanism called an injector removes a specific volume of clean water from the clean water tank, a specific volume of treated water from the treated water tank, and a specific amount of fertilizer from the mixing tanks and blends all three of these solutions together. The final blend of nutrient water is then pumped out to our plants in the greenhouse!

Water Distribution & Collection

When the nutrient water is distributed to the crops, it is fed directly to the root system of each plant via drip irrigation tubes.

Although every plant is fed the right amount of nutrient water, it is straightforward for the water to build up in a growing medium and potentially cause root damage – unless the excess water is drained away. For this reason, you can use coconut fiber as a growing medium because it drains excess water well, keeping the roots of our plants safe!

Below every row of plants, build gutters that catch excess nutrient water. Gutters should be built on a slight slant, allowing the drainage water to be gravitationally fed into the collection basin at the lowest point of the greenhouse. From here, the drainage water is pumped into the untreated water storage tank until it is cleaned and added to the treated water supply.

Closed-loop irrigation system enables us to irrigate efficiently, accurately and sustainably!

Benefits of Greenhouse Irrigation System

All modern irrigation systems are useful in several ways, depending on how you use them. Here few reasons why you should consider setting up a Greenhouse Irrigation System.

• *Filtration Systems*

Most greenhouse irrigation systems use filters to prevent small waterborne particles from clogging the small emitter path. New technologies that reduce clogging are now being implemented. Some domestic systems are installed without additional filters – as drinking water is already treated at the water treatment plant.

Almost all greenhouse equipment companies suggest that filters should be used in the system. In addition to other filters in the overall network, last line filters are highly recommended just before the final distribution pipe due to sediment settlement and unintended accumulation of particles in the middle lines.

• *Water Conservation*

Greenhouse irrigation can ensure the conservation of water by reducing evaporation and deep drainage compared to different irrigation systems, such as flood irrigation or overhead sprinkler irrigation, as water can be applied more precisely to plant roots.

Also, drip can remove many diseases that spread through contact with foliage with water. In areas where water supply is limited, there may be no actual water savings. Still, in desert areas or sandy soils, the system will provide drip irrigation flows as slowly as possible.

• Working and Efficiency Factors

Drip Irrigation, also known as trickle irrigation, works by gradually and directly supplying water to the root of the plant. The high performance of the system is due to two key factors.

They draw water into the soil until it can evaporate or run out.

It applies only to the water where it is needed, that is to say, to the roots of the plant rather than everywhere else. Drip systems are simple and relatively forgiving in design and installation errors.

It is a very efficient system for watering plants. For example, the typical sprinkler system has an efficiency of approximately 75-85 percent. The Greenhouse Irrigation System, on the other hand, has an efficiency level of over 90%. Over time, this difference in water supply and productivity can make a significant difference in the quality and bottom line of crop production.

In areas where water is scarcely available, such as desert areas of the world, the Greenhouse Irrigation System has unsurprisingly become the preferred irrigation process. They are relatively inexpensive and easy to mount, compact in nature, and help improve plant health at optimum moisture levels.

• Cost-Efficient

Irrigation systems are essential for modern agriculture as they significantly increase crop production. The Greenhouse Irrigation System may seem expensive in the short term, but it will save you money and effort in the long run. For example, this system can help lower production costs by at least 30% because you control the amount of water, agrochemicals and labor costs. However, it is advisable to have a quality greenhouse irrigation system with significant benefits.

Chapter 9: Greenhouse Heating, Cooling and Ventilation

Cycle Greenhouses should provide a regulated environment for the growth of plants with adequate sunshine, temperature and humidity. Greenhouses need to be exposed to maximum light, especially in the morning hours. Analyze the location of existing trees and buildings when choosing a greenhouse site. Water, fuel and electricity allow environmental controls that are necessary for favorable results to be carried out. For this purpose, make use of efficient heating, cooling and ventilation. Warning devices may be desirable for use in the event of power failure or extreme temperatures.

The house temperature requirements depend on which plants are to be grown. Most plants require daytime temperatures of 70 to 80 degrees F, with nighttime temperatures slightly lower. Relative humidity can also need some regulation, depending on the plants cultivated.

Some plants grow very well in cool greenhouses with night temperatures of 50 degrees F after being transplanted from the seedling tray, which include azalea, daisy, carnation, aster, beet, calendula, camellia, carrot, cineraria, cyclamen, cymbidium orchid, lettuce, pansy, parsley, primrose, radish, snapdragon, sweet pea and several other bedding plants.

Some plants grow best in warm greenhouses at night temperatures of 65 degrees F. Which include rose, basil, poinsettia, lily, hyacinth, cattleya orchid, gloxinia, geranium, gardenia, daffodil, chrysanthemum, coleus, Christmas cactus, calla, caladium, begonium, African violet, amaryllis.

Tropical plants typically grow best in high humidity at night temperatures of 70 degrees F.

Heating

Greenhouses must be heated for the growth of crops throughout the year. A good heating system is among the essential steps towards productive plant growth. Any heating system that offers a consistent temperature control without releasing material that is hazardous to plants is appropriate. Suitable sources of energy include natural gas, LP gas, fuel oil, wood and electricity. The cost and availability of these sources can vary slightly from one region to another. Convenience, expenditure and operating costs are all additional factors. Labor savings could justify a more costly heating system with automatic controls.

The specifications of the Greenhouse heater depend on the amount of heat loss caused by the structure. Heat loss from the greenhouse usually occurs in all three forms of heat transfer: conduction, convection and radiation. Usually, many types of heat exchange occur at the same time. The heat demand for a greenhouse is usually calculated by combining all three losses as a vector in the heat loss equation.

Conduction

The heat is carried out either through a material or through direct physical contact between objects. The rate of conduction between two objects is dependent on the area, the length of the path, the difference in temperature and the physical properties of the substance(s) (such as density). Heat transfer by conduction is most quickly reduced by replacing a material that conducts heat quickly with a weak thermal conductor (isolator) or by putting an insulator in the heat flow direction. An example of this would be to replace the metal handle of the kitchen pan with a wooden handle or to isolate the metal handle by covering it with wood. Air is a very low heat conductor and therefore a good heat insulator.

Convection

The heat transfer is the physical movement of hot gas or liquid to a colder location. Heat losses due to convection inside the greenhouse are caused by airflow and infiltration (fans and air leaks).

Heat transfer by convection involves not only the movement of air but also the movement of water vapor. When water evaporates in the greenhouse, it absorbs energy. As water vapor is condensed back to a liquid, it releases energy. As a result, when water vapor condenses on the surface of the object, it releases energy to the outside environment.

Radiation

Thermal transfer occurs between two bodies without direct contact or the need for any medium such as air. Like light, heat radiation goes through a straight line and is either reflected, transmitted or absorbed when an object is hit. The radiant energy must be absorbed to be converted to heat.

All objects emit heat in all directions in the form of radiant energy. The rate of heat transfer of radiation varies with the area of the body and the temperature and surface characteristics of the two bodies involved.

The radiant heat loss of the object can be minimized by covering the object with a highly reflective, opaque shield. Such a barrier:

(1) Reflects the radiant energy to its source

(2) Absorbs minimal radiation so that it does not heat up and radiate energy back to outside objects

(3) Prevents objects from seeing each other as a necessary element for a radiant energy exchange.

Factors Affecting Heat Loss

Heat loss due to air infiltration depends on age, state and type of greenhouse. Older greenhouses or those in poor condition usually have cracks around doors or holes in the covering material through which vast amounts of cold air can enter. Greenhouses covered by large sheets of glazing materials, large sheets of fiberglass or a single or double layer of rigid or flexible plastic have smaller infiltration.

The greenhouse ventilation system also has a major influence on infiltration. Inlet and outlet fan shutters also allow a large air exchange if they are not tightly closed due to poor construction, debris, damage or lack of lubrication. Window vents seal well compared to inlet shutters, but even maintenance is important to ensure a good seal when closed.

Solar radiation enters the greenhouse and is absorbed by plants, soils and greenhouses. The warm objects then radiate this energy outward. The quantity of radiant heat loss depends on the form of glazing, ambient temperature and cloud cover. Rigid plastic and glass materials exhibit a "greenhouse effect" as they require less than 4 percent of the thermal radiation to pass back to the outside.

Heat Loss Measurements

Heat loss by conduction may be calculated using the following equation:

$$Q = A (T_i - T_o)/R$$

Q = Heat loss, BTU / hr

A = Area of greenhouse surface, sq ft

(Ti-To) = Air temperature differences between inside and outside

R = Resistance to heat flow

Minimum Design Temperatures

Good outside temperature for use in heater desi. Another requirement that the heater must fulfill is to provide enough heat to prevent plants from freezing during times of extremely low temperatures.

Other Heating System Design Considerations

Plastic greenhouses also build up moisture inside the structure, because there are almost no gaps or openings in the glass building. High humidity can lead to increased leaf and flower diseases. The forced air heating device helps to balance the air within the structure and helps avoid temperature change inside the room. Also, it is ideal to have fans running around the walls, combining warm air with cooler air near the floor. They can be worked continuously during cold times even when the heater is not on.

Duct systems for uniformly spreading the heated air from the forced hot air furnace are ideal. Two or more small heating units are more suitable than the larger unit because two units give more protection if one unit fails.

The alert tool is good protection if the heating system malfunctions or if there is a power outage. Some greenhouse operators prefer to have a battery-powered warning device to alert them if the temperature is below the appropriate range.

Ventilation

Ventilation lowers indoor temperature during warm days and provides carbon dioxide, which is essential to plant photosynthesis. Another benefit of ventilation is to eliminate humid, moist air and replace it with dryer air. High humidity is objectionable because it induces moisture condensation on cool surfaces and helps to increase the incidence of disease.

Manually operated roof ventilators ventilate few glasshouses. Usually, this approach is not sufficient for the ventilation of plastic shelters due to the risk of rapid temperature fluctuations. In Georgia, ventilation fans are highly recommended.

Winter ventilation should be planned to avoid cold swelling of plants. This was a problem with some systems using shutters at one end of the house and an exhaust fan at the other. The problem can be minimized by putting the intake high in the roof and using baffles to deflect the incoming air.

Free winter ventilation can be provided utilizing a convection tube system consisting of exhaust fans and fresh air inlets located in the roof and end walls. It is attached to a thin plastic tube stretching the length of the greenhouse. The tube is hung on a wire near the ridge and has holes over its entire length. The fans can be regulated thermostatically. The operation of the fan creates a small drop in air pressure within the greenhouse, which allows fresh air to flow into the inlet and to inflate the tank, which discharges air into the house through the holes in the tank. The holes emit "jets" of air which should be projected horizontally to ensure proper distribution and mixing with warm air before reaching the plants.

The thermostat prevents the ventilation when the desired temperature is reached; the tube breaks and the ventilation ceases. Within a closely built greenhouse, there is no difference where the ventilators are placed in the convection tube ventilation as the distribution of the air is determined by the tubes. In general, less fan capacity is needed for the convection tube system than for any other winter ventilation system. More air is required as the outside temperature rises to the point that the maximum capacity of the tube is reached. By this time, the outside air usually is warm enough to be entered through doors or other openings at the plant level.

Fans can be attached to or possibly combined with a cooling pad for use in evaporative cooling. In reality, air can be drawn through the pad with or without water in the pad. During warm times, enough air needs to be extracted from the house to provide a full air exchange every 60 seconds. Control fans with thermostat or humidistat to ensure proper temperature and humidity.

Greenhouses that are equipped with an evaporative cooling pad system with three or fewer fans should have one fan with a two-speed engine to prevent excessive temperature fluctuations and fan cycling. Select all fans to work under slight pressure (1/8 inch static water pressure). Fans not rated against low pressure usually move just 60 to 70 percent of the allowed airflow when built-in greenhouses. It is recommended that only fans tested and checked by an independent testing laboratory, such as AMCA, be used since this is the only guarantee that the product ventilation rate is achieved.

Exhaust Fans in End Wall

End wall fans are the most common method of forced ventilation. The air passes through the motorized shutter (winter), and the exhaust fans pull through the greenhouse.

The exhaust fans should be able to transfer small amounts of air without a draft (winter) and still have enough ventilation power for an air exchange inside the house every minute during the summer. One air exchange in a minute (without evaporative cooling) will maintain the temperature around 8 degrees F above the outside temperature. One-half of this volume of air will produce a temperature increase of about 15 degrees F. In contrast, two air exchanges per minute will cause a temperature increase of about 5 degrees F. Ideally the length of the house should not exceed 125 feet by this method. Houses up to 250 feet long, however, have been successfully ventilated using this process. Temperature differences are higher in more extended rooms, so higher ventilation levels are desirable. Only air must be permitted to enter the house on the sides or at the end of the fan.

Glazing in glass houses must be set up correctly and the houses in good condition to avoid a large amount of air from escaping into the building. If cooling pads are used throughout the summer, remove the motorized shutter and close it to prevent hot air from reaching the shutter and bypassing the cooling pads. You may attach a perforated plastic tube to the same inlet shutter to ensure proper air circulation for cold weather ventilation.

The same principle applies to multiple ridge houses, provided that each end wall is so equipped. A two-speed fan is normally used in a small hobby house.

The total inlet opening at the end of the wall for summer ventilation (shutter and evaporative pad vent) would provide about 1.5 square feet per 1,000 cubic feet per minute of air flowing through the working fans. A motorized shutter and one or two fans may be connected to one thermostat. In contrast, the remaining fans are connected to another thermostat, with air being supplied to these fans via the vent panel containing the evaporative pad.

Pressure Fans in End Walls

Window ventilation, which is 100 feet or less, can be carried out by installing pressure fans, which blast air into the building, high in the end walls.

The end wall fans are usually two-speed and powered by separate thermostats. To prevent high-speed air from hitting plants, a baffle is positioned in front of the fans to direct the air in the desired direction. Fans should have a protective hood to prevent the rain from blowing into the house.

The pressure fans are mounted in the sidewall of this device. The pressurized device with fans on the sidewall does not work well when the vegetation is thick, and a lot of tall, growing plants are present. Note that the air outlet and inlet are on the same side of the house in this situation, with a box enclosure around the fan where the cooling pads are located.

Evaporative Cooling

The heat absorbed on a deep surface perpendicular to the sun's rays can be about much as 300 BTU / HR per square foot of the air. Thus, theoretically, it would be possible for the greenhouse to consume 300 BTUs per hour per square foot of floor space. This excessive energy contributes to a build-up of heat and, in warm days, can cause plants to wilt.

Excessive heat build-up can also be avoided by shading materials such as wood roll-up windows, aluminum or vinyl plastics as well as paint-on materials (shading compounds). Roll-up panels, which fit well in hobby homes, are available with pulleys and rot-resistant nylon ropes. These displays can be changed from the outside as temperatures vary. With this process, radiation can be decreased by 50 per cent, which will minimize the temperature increase in proportion if the rate of ventilation remains constant. Shading also decreases the effect of light on plants, which can restrict their growth rate as light is necessary for photosynthesis. This is a trade-off that is often required to reduce temperatures.

When summer temperatures surpass those considered acceptable and cannot be adjusted at a reasonable rate of ventilation and shading, the only solution is evaporative cooling. The fan and pad system, using evaporative cooling, removes excess heat and restores moisture. This reduces plant moisture loss and therefore reduces plant wilting. The temperature is reduced, the humidity is increased, and the watering needs are decreased.

The evaporative cooling system pushes air through a panel or sprays water in such a way that water evaporation occurs. Approximately 1,000 BTUs of heat are needed to convert 1 pound of water from liquid to vapor. When the heat comes from the air for evaporation, the air is cooled. Evaporation is higher when the air entering the system is warm, i.e. when the relative humidity is low, enabling the air to evaporate a lot of water. The water keeping power of the air is expressed in terms of relative humidity. For instance, the relative humidity of 50 percent means that the air retains one-half of the average water that the air could hold if it were saturated at a given temperature.

Theoretically, air can be cooled by evaporation until it reaches 100% relative humidity. Almost 85 percent of this temperature drop can be achieved by an excellent evaporative cooler.

Evaporative coolers are more effective when moisture is low. Fortunately, the relative humidity usually is weak during the warmest days of the day. Solar heat entering the house compensates for some of the cooling effects. A well-designed ventilation system providing one change in air volume per minute is necessary for a successful evaporative cooling system. With one change of air per minute, a solar heat gain of 8-10 degrees F can be expected. If the outside air was 90 degrees F and the relative humidity was 70 percent, the temperature inside the house will be around 93 degrees F.

If a cooling efficiency of 85 percent is to be achieved, at least 1 square foot of vertically mounted pad area (aspen fiber) should be given for each of the 150 CFM of air circulated by the ventilators. Many pad materials have been successfully used, provided the complete water film does not shape and impede the flow of air through the wet surface.

Aspen pads are typically limited to a welded wire mesh. A pipe with closely spaced holes allows the water to flow down a sheet metal spreader on the pads. Water, which does not evaporate in the air stream, is trapped in the gutter and returned to the river for recycling. The reservoir will have the capacity to carry the water back from the pad when the machine is turned off.

A cover of some kind is needed to prevent air from flowing through the pads during cold weather. They can be operated manually or automated. Float power is easy to regulate the water flow. It is desirable to use an algaecide in circulating water to prevent the growth of algae on the pads. You must also prevent rainwater from entering evaporative cooling water, which allows the chemical mixture to be diluted.

The evaporative pads in the suction side of the ventilators that discharge air into the house (pressure fans) have not worked well, mainly due to the distribution of cooled air. The same applies to package evaporative coolers, where poor air distribution is concerned. These units can handle air volumes from 2,000 to 20,000 CFM. The problem with them is the challenge of having a consistent distribution of cooled air. The closer the units are spaced around the walls, the greater the distribution of the climate. Package refrigerators have been used in small houses and houses with proper air distribution, with considerable success. The pressurized system forces air, which must displace the air inside the house, into the greenhouse. Vents must be provided for the circulation of air.

Mist Cooling

The evaporative cooling by spraying tiny droplets of water into the greenhouse has had minimal success. The droplets must be small, and this requires small, narrowly spaced nozzles worked at reasonably high pressures — a costly design. Water must be well filtered to avoid clogging of the nozzles. It is difficult to achieve a consistent distribution of water droplets in the room.

If the mist system contains any minerals in the water, the deposits will stay on the foliage of the plant. This accumulation can significantly reduce photosynthesis and can contribute to salt toxicity. The mist system can also cause wet foliage, leading to disease problems, especially when the size of the droplet is too large.

Mist cooling does not cool as efficiently as traditional evaporative cooling pad systems but is less costly. The system does not require a collection pan or a sump. It can cause runoff or pudding under the pads if all the water sprayed onto the pads is not vaporized.

A device which is a combination of a cooling pad and a misting (or fogging) device is known as a fogging pad system. Some growers have been using it with success.

Approximately 20 gallons of water per minute to be sprayed on the pad (usually 20, 1-GPM spray nozzles) for each 48-inch fan in the ventilation system should be installed. However, this amount of water will not always be needed.

Warmer air evaporates water faster than cooler air. The amount of water applied to the pads can be changed using a combination of valves, clocks and thermostats. As the temperature in the greenhouse rises, so does the frequency of activity of the mist nozzle.

Natural Ventilation

Some greenhouses can be ventilated by the use of side and ridge vents, which run the entire length of the house and can be opened as needed to provide the desired temperature. This approach uses thermal gradients to establish circulation due to the rise of warm air.

Houses with only side vents depend on the wind pressure for ventilation and are generally not satisfactory. Warm air must be enabled to rise through the ridge vent while cooler air is entering along the sides. The size of the vent is significant. Ridge vents should be about one-fourth of the floor area, and the side vents should be about the same amount. The roof vents should be opened above the horizontal position to provide a 60-degree angle to the roof. Most of these vents are operated manually.

Chapter 10: Best Greenhouse Equipment and Accessories

Efficient greenhouse equipment and accessories can be critical to the successful operation of the greenhouse.

To be able to grow high-quality, year-round crops, you would need to know what kind of greenhouse equipment you need to buy.

As indoor growing is emerging to be standard practice for growers around the globe, your greenhouse equipment and accessories will need to fit into tight spaces and mini-greenhouses.

Let's look at the best greenhouse equipment and supplies required when beginning greenhouse gardening:

Best Greenhouse Equipment You Need

One of the main features of an effective greenhouse design is getting the right accessories to make growing crops easier.

Type of Greenhouse Equipment

1. Simple
2. Greenhouse Furniture
3. Water Management — Irrigation and Drainage
4. Lighting

5. Heating and Climate control
6. Ventilation
7. Pest Control

Each of these components must be prepared for the design of the greenhouses.

As you read on, we're going to explore the various things that come in each group.

Basic Greenhouse Equipment

These are the fundamental things that you need to start growing plants.

Your choice of greenhouse containers is essential because it will have a significant influence on how your vegetables or plants grow.

Gardeners can use almost anything that grips the soil as long as it meets two criteria for a greenhouse container:

First, it should promote good health, provide plenty of rooting space and provide excellent drainage.

Second, it should keep the crop well and stabilize its upward production.

There are various kinds of containers, such as flats and tubes, hanging baskets and pans.

There are also larger containers that are designed to accommodate a variety of smaller pots.

Hanging baskets are ideal for growing plants, flowers and vegetables in height while making good use of space.

It can be made of plastic, metal-ceramic or even coconut fiber.

Plugs and flat containers are also used for early germination purposes.

These containers are available to accommodate several small plants or flowers while keeping them apart.

As far as potting is concerned, gardeners prefer greenhouse pots made of clay because they are the traditional way to grow flowers and plants.

However, if you cannot buy clay pots, you can also consider materials such as plastic, wood, peat moss and wood fiber.

They are also lighter in weight, more durable and cheaper than clay pots.

They're also easily disposable.

Seed Boxes

Seed boxes are also considered as essential greenhouse equipment.

Plastic seed boxes take over from the wooden boxes used by many growers in the past.

The pros and cons of plastic against wood are still being explored across the world.

If you want to go green, cut off the plastic option and stick with the wood.

However, get seed boxes that are around 14 inches X 8 inches X 2 inches.

This is the ideal size for rising baby seeds.

Rain Chains like these copper rain chains are another perfect product touch.

Choosing Good Greenhouse Pots and Containers:

The soil is porous greenhouse containers will dry out quickly, and you'll have to water it many times, so you'll have to wastewater.

For non-porous containers, the soil tends to retain moisture better, while preventing over-watering.

When choosing greenhouse pots and containers, bear in mind that they should do more than satisfy the growing needs of a plant; they should also have good drainage and porosity.

If you plan for year-round crops, you will need to prepare for mobility.

Your greenhouse pots should be easy to push and lightweight.

You'll need to make sure that your containers are environmentally friendly, which, in effect, will provide the ideal atmosphere for your plant.

Right greenhouse pots and supplies will raise baby plants from seed even if the soil is not consistent with germination.

Choosing the best type of greenhouse container will make a significant difference in the overall growth of your greenhouse plants.

Furniture to Store Your Greenhouse Equipment and Plants

Well-planned furniture and adequate shelving inside a greenhouse are essential for the storage of all your pots and containers.

In a tight, space-limited greenhouse system, shelving will improve the growing area without negative impacts on the shade.

Some greenhouse shelves are mobile (with wheels) and can be moved outdoors during ideal weather and back indoors at night or during cold temperatures.

Shelves may be made from materials such as glass, wood or metal.

It is necessary to note that the amount of lighting entering plants can be affected if double shelving is used.

You can also find shelves under garden benches to save space.

Greenhouse shelves may be temporary for starting seedlings or permanently attached to the greenhouse structure.

Cinder, wooden blocks and metal are suitable for legs and tables.

The wire mesh of the shelves makes it possible to remove excess water.

Greenhouse shelving can also help keep crops apart to avoid seeding or cross-pollination.

Garden benches are one type of shelving that allows maximum room and storage.

In general, their ideal size is determined by the width of the hothouse to maximize the through space.

Benches can be permanent or temporary fixtures.

If you plan to delete or rearrange them regularly, then sectioned bench options can be ideal for you.

Planters are another kind of greenhouse furniture that is commonly used in today's gardening environment.

Large and deep seedlings are generally recommended for food crops.

They can be made from a range of materials, such as plastic, wood or metal.

They are usually organized in such a way that each planter contains only one vegetable.

Greenhouse Irrigation and Drainage

Systems over their lifetimes, you will need ways to water your plants.

Though automatic watering systems are all crazy, there will always be a special place in traditional gardening for good old watering cans.

The long beams on the cans will comfortably touch all the plants, including at the back of the flower bed.

Greenhouse Water Management

You will likewise be able to customize your watering experience depending on what each bed needs.

Plastics also take over metal in the watering can section.

Plastic containers are usually smaller, making watering less labor-intensive.

Plus, they're also cheaper.

But if you have greenhouse aesthetics, stick to the metal cans.

The trickle watering system is another form of greenhouse equipment.

Greenhouse Trickle Watering System

Build a plastic hose rinse system with outlet nozzles at various intervals across its duration.

Place the hose around your pots in acceptable proximity.

Attach your hose to a storage tank that keeps filling and releases water when it's finished.

What's great about this system is that you're going to water your pots or beds in the exact quantity you want, at the exact time every day!

Other types of greenhouse water treatment equipment that you may need are good pumps, water breakers, valves, pumps, hoses, sprinklers and temperature control boilers.

Remember, you're going to have to find a place in your greenhouse or yard to store these things.

Greenhouse Lighting

The lighting system inside the greenhouse determines the level of sunlight, artificial light and shade of the plants.

If the sunlight in your area isn't strong enough, you might need to consider artificial lighting.

- Grow lights
- Seedling lights
- LEDs
- General all-purpose lighting
- High-intensity lights

For providing a vast lighting network can be costly for small greenhouses, for larger ones, it is almost necessary.

Climate Control and Heating Greenhouse Equipment

These are often a few types of greenhouse equipment designed to control the amount of moisture, heat and frost inside the greenhouse.

Let's look at the main components of the climate control system:

Greenhouse Thermometers

The key to a successful greenhouse is to maintain perfect temperatures at all times, making the thermometer very important.

Install a maximum and minimum temperature thermometer with a needle location that appears when the mercury is removed.

When it comes to greenhouse thermometers, you're going to need one that can be reset with a magnet.

There are more high-end options with push-button readjustments, but they are not necessary.

All types of greenhouses, even portable ones, require thermometers.

Often, call a soil thermometer to determine the temperature of the soil.

Greenhouse Thermostat

A greenhouse thermostat helps you to know the current temperature in your greenhouse and to control it accordingly.

The temperature gauge or thermometer indicates the temperature changes while the thermostat automatically controls the temperature in the desired area.

Ideally, a good greenhouse is expected to have a thermostat.

Greenhouse Heaters

Other critical greenhouse equipment deals with the control of heat inside the greenhouse.

Greenhouse heaters are needed to control the temperature of the greenhouse.

They come in different types, with different modes and sources of energy.

You can choose from electric, gas and propane heaters according to your needs and requirements.

You also have the choice of either choosing a sold or non-vented heater.

Greenhouse Humidistat

Greenhouse humidifiers or humidistats are needed to check the quantity of moisture in the greenhouse.

Some plants are sensitive to dry air, which will impede their vegetative growth.

With the aid of an appropriate humidistat, this problem can easily be solved.

Ventilation Equipment for Your Greenhouse

Proper ventilation is compulsory for adequate plant growth, not only during certain seasons but throughout the year.

This is because, at any time of year, the sun is capable of causing extreme temperature changes.

A reasonable rule of thumb is to have open venting options equal to around 20% of the floor space.

Vents can be found on the roof and sides of the structure as well as part of the entrance.

Roof venting is considered to be the absolute best when it comes to fixing venting systems.

Many automated venting systems are the perfect choices for those who are not around to handle the greenhouse all day.

Exhaust fans are another way to vent excess air, but whether or not they are a good option for any case, in particular, they deserve analysis.

Pest Control Equipment

No list of greenhouse equipment will be full without pest control equipment.

There are several methodologies for effective pest control; some use chemicals and others use biology.

Chemicals are easy to use and relatively cheap, but others may argue that they do more harm than good.

Natural methods, such as the use of what is generally referred to as "beneficial insects," are another type of pest control.

Mostly, these bugs track down and eat the bugs that kill your garden.

Often all the plant wants is a decent mesh to keep the pests out.

These meshes may be made of metal, fabric or thin plastic.

You would also need fencing and door sweeps to keep bugs out of particular areas.

Various kinds of fogging machines and sprayers kill bugs.

Insecticides and pesticides should be used sparingly, ideally using organic or natural sprays to prevent pests from damaging crops.

Greenhouse Equipment and Accessories Soil Sterilizers

Whatever soil you are considering for planting, it would be extremely helpful to make use of a sterilizer.

There are several ways to sterilize the soil, but the easiest and most effective way to do so is by using a steam sterilization system.

Steam systems are advantageous and inexpensive.

They're not going to take up a lot of space and do a fantastic job on your soil.

Gardening Sieve – Sowing Sieve

The soil texture is an important consideration when planting baby plants.

A sowing sieve will be very useful in helping you to achieve the perfect texture.

You can use a mesh sieve to cover your seeds with compost after you plant them gently.

The good news is that you don't necessarily need to buy a sieve.

This piece of greenhouse equipment could be a DIY idea.

Use a small wooden box like those in which you buy bulk produce.

Take the bottom of the box and put a piece of perforated zinc in it, and voila — you've got your sieve!

You can also assign this essential task to your children and your relatives.

Plant Support Equipment

You will need to provide adequate support to ensure that your plants grow in strength and length.

Sometimes, all you have to do is tie the plants together so that they can support each other.

Although there are many materials used to bind your plants, we recommend raphia as it is moderately priced and can help most plants.

Many greenhouse growers often use broken loops, green growing twines and paper-covered wires.

Fencing and greenhouse molds can also be used to shape plants into the desired shapes.

Chapter 11: Greenhouse Vs Polytunnel

How to Decide?

If you've been growing your fruit and vegetables and are looking for ways to improve your crop, you might consider investing in a poly-tunnel or greenhouse. Providing protection against harsh weather conditions and many pests, these could be perfect for extending the growing season, but which alternative do you choose?

This guide discusses the distinctions between poly-tunnels and greenhouses, as well as several factors that need to be considered when making your decision. Do you want to know which one is best for both your growing needs and your garden? Read on to find out about it!

Poly-Tunnels and Plastic Greenhouses

What Is the Difference?

Essentially, both systems function similarly, providing a warmer climate to promote growth during the summer and allow low-temperature plants to survive in the winter. This will help you grow a greater variety of crops (including more Mediterranean crops), in larger numbers and over a more extended period. But what are the significant differences?

Poly-tunnels are made of galvanized steel hoops and are tightly linked with transparent or diffused material. Cheaper to buy, poly-tunnels require a smaller investment; at only one-third of the cost of a greenhouse, poly-tunnels are excellent value for money. Poly-tunnels are typically constructed directly on the base of the soil, with a simple site clearance necessary for the preparation of the site. They're also easier to dismantle, move and reassemble. The galvanized steel frame will last for 20 years and more, and polythene generally needs to be replaced every 7 to 10 years. Poly-tunnels are available with a full range of additions and accessories to allow you to customize your poly-tunnel to meet the specific needs of you and your plants.

Greenhouses uses an aluminum or metal frame made of glass or polycarbonate plastic. This can be costly, particularly if you choose toughened safety glass (recommended for family gardens or homes with young children). Greenhouses are often time-consuming to build, usually requiring a firm foundation to stand on, and once built, they are difficult to disassemble. On the other hand, they are hard-wearing and durable.

Whatever you choose, options may be ideal, especially if you are new to growing your fruit and vegetables and are interested in becoming more self-sufficient.

A small poly-tunnel or greenhouse can be assembled relatively quickly, it is perfect for growing popular plants such as tomatoes, and can be a great way to get kids into gardening. Also, they are a perfect choice for smaller outdoor areas, from yards to crowded city gardens.

Big Poly-Tunnel or Large Plastic Greenhouse

A large poly-tunnel or large plastic greenhouse is an excellent choice for those who need more space whether it's because you're a more experienced gardener and want to increase your crop variety or require extra space for a commercial venture.

Whatever the cause, a larger size will allow you to take your growth to the next level, both for personal use and professional use.

Factors to Consider

When choosing between these two structures and selecting the scale, consider the following considerations:

1. *Purchase Price*: Poly-tunnels are usually cheaper and need less investment than greenhouses.

2. *Construction Time*: Depending on the scale, it can take longer to create the greenhouses.

3. *Site Preparation*: The soil must be level for a greenhouse, while the rougher ground can be more easily accommodated for a poly-tunnel.

4. *Lifespan*: Greenhouses will theoretically last a lifetime, given that the glass does not break or blow out of the frame. At the same time, poly-tunnel covers may need to be replaced regularly, at a low cost, to preserve their performance.

5. *Heat Retention*: Greenhouses also need more heat during the winter months, while a poly-tunnel covered with thermal polythene and without a greenhouse design may help if you consider over-wintering crops.

6. _Light and Shades_: Diffused polythene sheeting on a poly-tunnel helps prevent heat spots, but you can need to paint greenhouse panels to avoid sun damage, such as leaf scorch.

7. _Transportability_: While both can be moved, greenhouses take more time, and glass panels may pose a higher risk of damage and injury.

Essentially, your preference will be based on your budget and specifications. However, we hope that these tips will help you choose the right through choice for your needs.

Chapter 12: Insect Management, Pest Control and Fertilization

The effect of plant pests on the aspiring greenhouse vegetable producer is clear and essential. The prospective producer must understand that Florida is a paradise for both the crop and its accompanying pests. Disease-causing organisms, insects, and nematodes can cause severe problems in greenhouses. Without a real wintertime, the pest population continues to grow, and many are managed throughout the year. As a result of this mild climate, the adaptability of both temperate and tropical pests to Florida presents a large number of potential greenhouse crop problems.

If a person is interested in building or starting a greenhouse operation, some important considerations need to be understood regarding the pests and their potential to reduce or destroy a greenhouse crop. First, a greenhouse provides a protected environment for pests to thrive. The essence of the idea of greenhouse farming is to grow plants by providing a means of shielding the crop from extreme heat, cold, rain and various environmental factors that would otherwise slow down or discourage the crop from growing at a given time. As a result, pests that may inhabit a greenhouse are shielded from the same harsh environmental factors that usually lead to their control when a crop is grown out of doors or under field conditions. For example, the driving force of rainfall and wind also helps to keep mites, aphids, and other insects under economic control. Direct sunlight and constant temperature changes often play an essential role in the overall natural pest control frequently accomplished under clear conditions. The greenhouse practically protects the plants and therefore, their respective pests from these environmental conditions. Nevertheless, biological control of insects may be more effective in a protected environment.

One serious problem to address is the lack of available greenhouse gas control measures. For example, tomatoes grown under field conditions may have several hundred insecticides and fungicides that can be used on them, not in the greenhouse. Strict EPA, federal and state law limits the number of pesticides that can be used on greenhouse tomatoes or other vegetables. A plant should not be used if it is explicitly labeled "not for greenhouse use" with instructions for a particular crop.

There are significant legal and liability issues that may be encountered by a greenhouse grower by using hazardous chemicals in enclosed areas. This is one of the main reasons why the majority of pesticides used under field conditions cannot be used in greenhouse situations. When comparing greenhouse with field use, pesticides can dry on the plant at different rates, become highly volatile and release toxic vapors longer or faster under greenhouse conditions, as well as react differently in many other chemical and physical ways. Toxic fumes cannot escape and are diluted as easily as they would under field conditions, thereby posing a potential risk to the grower. There are several considerations that growers must make when using pesticides in a situation where staff may be exposed daily and regularly under the changed and enclosed conditions of the greenhouse.

The greenhouse grower also faces the question of repeated harvesting over a long period. Greenhouse tomatoes, for example, are harvested many times, and a larger number of harvests often prevent the use of many effective pesticides with a longer pre-harvest time limit. The advantage of growing vegetables under greenhouse conditions is the ability to pick up small amounts of ripe fruit every day or so, rather than harvesting large quantity in a minimum number of harvests. Development characteristics of greenhouse fruits require close and careful handling by harvesters, and pesticide residues are again a potential problem under these conditions. Another problem that a potential greenhouse grower must face is that the equipment available for the efficient use of pesticides under greenhouse conditions can be minimal. For example, the field tomato grower has a wide range of application equipment options, such as aircraft, sophisticated tractor equipment and chemistry units for use in their operations. Greenhouse growers often only have access to hand or small capacity power units that are severely limited in nozzles pressure and spray application versatility. The lack of proper and usable equipment makes it difficult to manage many pests under greenhouse conditions. To use scarce and costly space, the greenhouse grower is often unable to use large pieces of spray equipment that are needed to perform a specific task. The grower must, therefore, be extra careful in the long-term planning, collection, use and application of pesticides.

Insect Management

Due to all the problems associated with the use of chemicals in the greenhouse climate, farmers need to use exclusion as their first line of defense. Insect management, therefore, needs to be regarded when building the greenhouse. Insect-proof screening is open for vents and other openings, but due to increased resistance to airflow, the surface area of the screened areas must be increased to account for this. For greenhouses filled in plastic, the use of ultraviolet-absorbing plastics can reduce insect problems. UV-free indoor light changes insect landing and feeding behavior. It can dramatically reduce the spread of insect-vectored viruses as well as deter the production of aphids, whiteflies and thrips.

UV-reflective mulches used on the ground near the greenhouse are helpful to limit the entry of these pests.

The protected greenhouse ecosystem promotes the survival of beneficial insects as well as pests, so biological control is another technique worth exploring. Much of the work on how best to use natural enemies in the growth of greenhouse vegetables is still ongoing. Still, growers are already experimenting with the use of lady beetles, lace larvae, pirate bugs, and predacious mites. This approach could be expensive. Predatory mites are particularly useful for controlling spider mites under greenhouse conditions. Suppliers of natural enemies may suggest the release of the proper species.

Nematode Management

In addition to conventional insects and diseases in greenhouses, nematodes can pose additional problems. Nematodes are among the pest/disease issues that can be especially troublesome in greenhouse systems. These relatively microscopic worms feed on or in the roots of plants, disrupting plant root growth and function. Some redirect significant amounts of plant energy to their growth and support, reduce quantity and yield quality and often delay crop maturity. They grow well at 80–90 ° F and cause significant problems for many of the crops most common in greenhouse systems. Nematodes are easily spread in contaminated water, soil or through media, and plant tissue. They can be particularly problematic in greenhouse systems because among other things:

• Nematodes are easily introduced into greenhouse operations and are very difficult to get out of them. Any of many dangerous plant nematodes are commonly found in most natural soils. They will invade a greenhouse crop whenever the barrier between the crop and the native soil beneath the house is breached. A single root that penetrates through plastic, concrete or other flooring material can provide a route of entry. Infested transplants are another particularly conventional means of putting nematodes in a greenhouse. Any part of the growing medium that has never been sterilized or exposed to contamination during storage or handling can produce nematodes, even water, if it is obtained from a shallow or surface source, can carry nematodes to the greenhouse.

• Nematodes are aquatic animals so they can be transmitted particularly easily in water. Hydroponic recirculation systems are vulnerable to nematodes, as the entire system can become uniformly infested as soon as nematodes in or on roots anywhere in the system begin to replicate and enter the water. Of course, any movement of soil or roots from the infested area to other parts of the house is also likely to cause nematode infestation.

• Once a greenhouse crop has been infested, there is no pesticide or other medication that can cure the problem in that crop. The problem can only be eliminated by destroying the crop and sterilizing every component of the growing system that has been contaminated. Since this is very difficult, perhaps almost impossible, in some systems, careful sanitation to prevent the introduction of nematodes into operation is by far the best means of nematode management.

Disease Management

The major greenhouse crops in Florida currently include cucumbers, cabbage, tomatoes and various herbs. As an indication of the extent of the pest situation, the following is an estimated list of the number of diseases that could potentially affect these crops in the greenhouse:

• Cucumber: 9 fungal and viral diseases

- Lettuce: 7 bacterial, fungal, and viral diseases

- Tomatoes: 21 bacterial, fungal, and viral diseases

- Herbs — various plant-specific diseases

Some of these diseases are yearly issues, and others appear with less frequency depending upon the season, variety, production requirements, etc. For example, hydroponic production systems can have whole crops affected by root or vascular wilt diseases as the root zone is continuously in the nutrient film. There are no legitimate disease control measures for either the herb or the lettuce production in the greenhouse and only one for the production of cucumber. The production of tomatoes has access to seven separate products for collective use for just ten diseases. None of these seven products is successful against diseases caused by bacteria or viruses. Moreover, due to the nature of these facilities, the usual field use of multipurpose fumigants for the control of soilborne pathogens is not adaptable to greenhouse growth. Hydroponic or field systems where root or wilt diseases exist are often challenging to disinfect.

To successfully control plant diseases in greenhouses, the method of crop production needs to be closely linked to disease goals and insect control practices. Insects, such as thrips and whiteflies, are vectors for viral diseases.

Pest Control Considerations

The planning stage for the production system must incorporate the following considerations:

a) Greenhouse design (especially height, heating, insect screens and ventilation components) and an irrigation system that minimizes leaf moisture and moisture at the plant canopy level.

(b) Range of viable pest-resistant varieties.

(c) Pest-free and safe transplantation minimizes the introduction of plant pests, insects and nematodes.

(d) Optimum fertilizer programs that ensures healthy growth rather than maximum growth.

e) Scouting of viruses, nematodes and insects during the growing season.

(f) Sanitation practices that minimize the movement of microorganisms from diseased plants to healthy ones, including removal of all plant material after final harvest.

(g) Harvest and shipping activities that optimize product quality.

The introduction of these integrated practices would ensure that greenhouse vegetable crops are economically and environmentally acceptable.

Chapter 13: Cleaning and Maintenance of Greenhouses

Like our houses, the greenhouses require daily maintenance to keep them in good shape and to maintain the required standards of cleanliness and hygiene. This comes down to looking after the layout of the greenhouse – within and outside – its heating, ventilation and irrigation systems, and maybe most importantly, the growing ecosystem itself.

General Maintenance

The greenhouse should be checked regularly for any broken or cracked glass, particularly after high winds, and the panels should be replaced as required. The frame, too, needs to be inspected routinely to make sure it is in good order. Although aluminum enclosures are usually reasonably low-maintenance, wooden enclosures will need to be regularly painted or treated with an acceptable preservative – making sure that none of them touches the plants, of course – and any rotting areas will need to be repaired.

Cleaning

Regular cleaning helps to maintain the structure of the greenhouse and dirty windows reduces the level of light – which can lead to problems with propagation, such as tight seedlings. The best times to have a big clean-up are in the spring – just before you start sowing – or in the early autumn before half-hardy plants are brought in for the winter.

Cleaning the exterior of the greenhouse is best achieved on a relatively breezy day, using warm water and a sponge, allowing the wind to dry the glass and stop leaving too much in the way of streaks. Aluminum framed greenhouses can be particularly prone to the collection of grime beneath the joints between the panels – the jet-wash attachment for the hose or the gentle scraping with the old plant label often shifts the dirt very successfully.

Regular, thorough cleaning of the interior of the greenhouse is essential, not only to create a pleasant environment in which to work but also as a significant part of the control of pests and diseases. It has also been said that it will be difficult to overemphasize the value of cleanliness for greenhouse management – so doing a good job here will save hours of trouble!

When beginning, make sure that the electricity is switched off, that the heaters and other electrical appliances are unplugged and that the sockets are sealed. If the temperature outside is too cold, remove all plants, containers, pots and staging – protecting any tender or semi-hard plants with horticultural fleece or putting them inside a shed or garage while you're working. This is the perfect time to scrutinize each plant, to remove any sick or damaged leaves and to discard anything dead or dying.

All spent compost or growing bags should be disposed of, and the floor swept away to avoid fallen leaves and general trash, and then any beds weeded. Now that the greenhouse has been emptied, the glass, the pathways and any brickwork should be thoroughly cleaned with warm water and an appropriate disinfectant, allowing them to dry thoroughly before returning the evicted plants to their proper places.

Check the Equipment

The heating, ventilation and irrigation systems are essential to the greenhouse, and they should not be forgotten or ignored. Daily checks – and, where necessary, servicing – of these main items are crucial, whether you do so yourself or call outside support. At the same time, pots, containers, seating, spread tables, capillary mats and the like should also be routinely sterilized to popular the risk of pests and disease and any lighting systems inspected.

Maintaining the Growing Environment

The dry, humid atmosphere of the greenhouse is suitable for many pests and diseases that can spread depressively quickly when given the opportunity, but keeping the greenhouse clean and tidy will significantly reduce the risk that they will become a problem. Supporting those plants that need it – and tying them in their developing shoots – combined with a regime of pruning, picking, potting and pinching plants as required and a thorough weeding around the beds would help.

Providing some form of shade as necessary can also be of great benefit – preventing leaves that are scorched and the almost unavoidable resulting disease attacks that the weakened and damaged plant would experience. It is necessary to remember that the perfect growing conditions for plants provided by the greenhouse are just as tailor-made for any number of bacteria, fungi and other unwanted guests. Part of daily greenhouse maintenance also includes being always on the lookout for pests and diseases – such as aphids, red spider mites, mealy bugs, mildew and botrytis – and ready to handle them promptly.

Spring cleaning is just as crucial to our greenhouses as our homes. Still, in addition to a one-off annual cleaning, routine maintenance is essential to keep everything running smoothly. However, with a little attention to one or two key elements, keeping our greenhouses at its peak does not need to pose too much of a problem or take too much of our time.

As the summer slips gently into the autumn, and the days begin to get shorter and noticeably colder, there are a few jobs that need to be done around the greenhouse to get things ready for the coming winter.

Seasonal Maintenance

Autumn is a good time for a little routine maintenance of the greenhouse – cleaning and repairing the structure inside and outside, thoroughly disinfecting the staging and equipment and ensuring that the heating system is in good order before the temperature drops. This can be done as early as late August or the first week of September, especially if you have a lot of tender plants to protect – British weather being what it is, you can never be sure when the first frost comes, so it's just as good to be prepared!

It is a great idea to use this opportunity to clean and inspect gutters for leaks, either by replacing or repairing, as appropriate – leaky gutting can often cause wood to rot in wood-framed greenhouses and does little to improve aluminum ones either. It is worth keeping a close eye on gutters and downspouts through the autumn, particularly if there are trees in the area, as they can block leaves very quickly at this time of year.

As the autumn progresses and the light levels drop, any shades put up during the summer to protect the delicate leaves from scorching should now be removed. If your greenhouse is on the exposed site or you keep the plants particularly tender, it may be time to think about insulation.

Plant Care

Before nights turn too cold, young or frost-sensitive plants need to be brought in, and it is an excellent plan to check them at this point for any signs of pests and diseases to avoid any problems in the greenhouse. With the greenhouse now often at its most crowded, careful attention needs to be paid to the first signs of any ill-health – mildews and molds being a particular potential nuisance at this point in the gardener's calendar.

As the indoor temperature becomes cooler, it is critical that all watering is done carefully to avoid making perfect conditions for the grey mold (Botrytis) that attacks a wide range of plants and thrives in a cold, damp climate, thus aiming for a slightly dry atmosphere. Prompt treatment will also help, but any plant that has gone too far will – sadly – have to be discarded and burned to prevent the spread of the disease.

Even though the summer has just passed, autumn is a good time to think about getting ready for the next growing season. A wide variety of different plants can be propagated from cuttings and overwintered in the greenhouse to give them the start of the New Year, including several tender flowering plants, shrubs, spices, carnations, fuchsias and pelargoniums.

Sown in early September, annuals can also be grown in the greenhouse – provided they get enough light – to provide a very colorful early show in the spring. Again, there is no shortage of suitable candidates, including calendulas, carnations, cornflowers, nemesis, godets, phlox and schizanthus. Early vegetables can also be sown, and bulbs can be planted in October to offer a welcome splash of color as the winter draws to a close.

There's always something a little sad about putting the garden to bed as autumn gradually starts to give way to winter – but at least the greenhouse not only allows you to shield your choice of tender plants from the worst of the cold but also to get ahead of the game for the coming season. With all your autumn maintenance and cleaning done, you will have just about enough time to pour through all those seed catalogues before you need to start sowing!

Chapter 14: Guide to Plant Herbs and Vegetables in A Greenhouse

A greenhouse can prolong your growing season and allow you to enjoy home-grown vegetables, fruit and herbs for a much larger part of the year, or it can be heated to allow for year-round growth. Greenhouses can be self-contained structures built on a sunny spot or small enclosures built adjacent to an existing outdoor wall. Whatever design you choose, be mindful that growing vegetables inside a greenhouse is different from growing them outdoors. Be ready to give your greenhouse plants a little more attention than you would if they were growing outside.

Growing Medium

Gardeners with well-drained soil can often create their greenhouse without a floor and use the soil nature-given. Fertilizers and modifications may be applied as necessary. The tilled, modified soil may be sterilized with a steam treatment that sustains a temperature of 180 degrees Fahrenheit for at least 4 hours. This sterilization kills any weed seeds, insects or active organisms that may be present in the soil. Raised beds can also be constructed and filled with rich, fertile soil if desired or if it is not possible to modify the existing soil. Avoid the temptation to cultivate greenhouse vegetables hydroponically unless you are a highly experienced gardener because these systems are high-maintenance and unforgiving.

Temperature Control

Greenhouses are cooled by air vents, shade cloths, screens and vinyl nets. Installation of an evaporative cooler may also be required. Choose an evaporative cooler that is rated 1 to 1.5 times the volume of your greenhouse for the best results. Heat is best provided by a professionally designed and adequately ventilated heater, whether powered by gas, oil or wood. Electric space heaters are also useful. It is normally not important to replace the greenhouse in your home's heating system or to mount a solar heating system cost-effectively.

Cold seasonal vegetables such as beets (Beta vulgaris), broccoli (Brassica oleracea), lettuce (Lactuca sativa), peas (Pisum sativum) and spinach (Spinacia oleracea) typically need daytime temperatures between 50 and 70 degrees F and nighttime temperatures between 45 and 55 degrees F. Cucumbers (Cucumis sativus), warm-season beans (Phaseolus vulgaris), peppers (Capsicum), eggplants (Solanum melongena) and tomatoes (Solanum Lycopersicum) favours daytime temperatures between 60 and 85 degrees F and nighttime temperatures between 55 and 65 degrees F. Herbs usually tend to live at night temperatures between 55 and 60 degrees F and daytime temperatures below 85 degrees F. However, some herbs, such as rosemary (Rosmarinus officinalis) and basil (Ocimum basilicum), require warmer nighttime temperatures between 65- and 70-degrees F.

Pests and Diseases

Due to the enclosed space and the close quarters, any disease or insect issues must be dealt with immediately in the greenhouse. Practice proper hygiene to prevent problems. Use only high-quality seed, clean tools with a 5-per cent bleach solution before planting seedlings, and avoid putting outside dirt in the greenhouse through shoes and garden tools. Maintain the humidity level as low as possible and ensure good air circulation in the greenhouse. After harvesting, remove the spent plants and hose the inside surfaces of the greenhouse as well as all the tools and equipment. If you find a problem, apply fungicides and insecticides quickly. Make use only products that are safe to use in plants intended for consumption, and always follow label directions.

Pollination

Some vegetables, such as seedless cucumbers, don't rely on pollination insects and can be grown in a greenhouse with no special attention. Other plants, however, such as tomatoes, eggplants, peppers and cantaloupe (Cucumis melo cantalupensis) rely on bees or wind for pollination. Since these two pollinators are not in the greenhouse, some of your greenhouse plants may need to be pollinated by hand. Plants that use the wind for pollination can be gently shaken before pollen is released from the male part of the plant and transferred to the female part. Some plants allow you to remove a male flower from the plant and rub it against a female flower. You may also use a small brush to move the pollen from one part of the plant to another. Male flowers rest on long, straight stems; female flowers have a bulge of undeveloped fruit on the stem beneath them. Your nearest garden center will be able to help you if you have trouble pollinating different plant varieties.

5 Most Common Vegetables and Fruits to Grow in the Greenhouse

There is a vast difference between the crops grown as starters and the greenhouse vegetables. Masses continue to grow vegetables and fruit in their greenhouses, where they regulate the temperature, provide heat, prolong the growing season and protect them from frosting. However, if you're new to gardening and planting fruit and vegetables, then you need to have good ideas about how to grow in a greenhouse. God has given us countless varieties, so how can you choose the best vegetables to grow in the greenhouse? Just start growing simple vegetables so that you can get your hands on them within a year so that the next season, you can keep growing even the complicated ones.

1. Leafy Greens: You have to start with something that belongs to the "salad family" – almost every other leafy vegetable grows in the same way, particularly when looking at bedding green house plants. Other than basic knowledge, some aesthetic knowledge is required when growing leafy vegetables. They have different tastes and colors, making them ideal for starters and sidelines. These can serve as a good source of income, as you can sell them to a variety of grocery stores and even wholesalers.

2. Micro Greens: In simpler terms, you will grow up with a lovely look and a mouth-watering taste of Tatsoi, Beet, Peas, Choi and Radish, etc. They are extremely loved as sidelines and as snacks. Once you have good knowledge, you can blend the varieties and make micro-greens of the second generation on your own.

3. Spinach: It is one of the best-grown greenhouse plants – if you want to enjoy the freshest and tastiest spinach, take it out of the garden and cook it immediately. It's so safe that you can quickly increase your intake of vitamins and minerals. Most importantly, you're never going to have trouble growing and preserving this greenhouse crop.

4. *Cucumber*: You must have grown up to eat cucumber salads or even raw bits of salt. They taste great – but growing them isn't so simple. You have to shrink wrap them so that they can preserve their freshness after harvesting.

5. *Tomatoes*: Many of the greenhouses have tomatoes in different colors and shapes – especially beefsteak varieties are easy to handle.

Chapter 15: Hydroponic in A Greenhouse

Nowadays, greenhouse growers are facing increased competition and increasing labor, energy and crop input costs. To increase their return on investment, farmers are slowly diversifying their crops to include hydroponic vegetables. Smart growers have realized that sustainable fruit and vegetable production can be a better investment and more productive use of time and money.

Changing patterns have expanded incentives for greenhouse growers to dramatically increase revenues and income using their existing facilities during the year. Today's hydroponic-growing methods have proved to make growth faster and more efficient than field-growing. Labor and crop input costs are lower, and quality is much higher. Converting greenhouses from conventional plant housing to edible products can be very simple and cost-effective. Almost limitless market prospects and low-cost investment mean that the future is bright for growers who want to grow hydroponically in established greenhouses.

Why Diversify?

The typical greenhouse grower is comfortable growing bedding plants, flowering potted plants, potted foliage plants and cut flowers. The regime that is best known to seasoned greenhouse devotees can be expensive and often financially challenging.

Many of these growers are now faced with increased competition and increasing labor, energy and crop input costs. To increase their return on investment, farmers are slowly diversifying their crops to include hydroponic vegetables—quickly finding that locally grown lettuce, tomatoes, cucumbers and peppers are in high demand.

Sustainable and locally grown food is a very hot topic. Many growers have realized that fruit and vegetables, grown in greenhouses all year round, are a good investment.

Big-budget customers, such as school districts and restaurant chains, are transitioning to locally grown food. States are growing the percentage of fresh produce that makes up school lunches, encouraging students and staff to live healthier lifestyles. Individual customers are increasingly interested in where their food comes from, and this curiosity will continue to rise and further increase competition. With transport costs increasing and food safety issues at all times high, the transportation of food by truck, ship and air has become prohibitive. With all these compounding problems, it should be clear that local is the way forward, but the farmers and consumers have still not been able to meet the growing demand for locally produced food.

The Hydroponic Advantage

These growing trends have expanded prospects for greenhouse growers to dramatically increase revenues and income using their existing facilities during the year. But the reluctance in which farmers add vegetables and fruit to their offerings is disconcerting. Pre-existing greenhouses can easily handle hydroponic growth with few changes.

Today's hydroponic-growing methods have proved to make growth faster and more efficient than field-growing. Labor and crop input costs are lower, and quality is much higher.

Hydroponic and greenhouse yields are usually ten times the yield of the field for a one-crop-per-year harvest. In some instances, hydroponic and greenhouse yields have been 100 times the yield of Bibb lettuce in the field.

Converting greenhouses from conventional plant housing to edible production is now very simple and cost-effective. Growers can turn their low-to medium-technology greenhouses into hydroponics without having to spend a large amount of money in a new greenhouse. Most farmers, with some work and patience, can handle the project on their own.

An increasing number of colleges and vocational schools have agricultural departments and curricula that appeal to potential students as committed, skilled growers. Banks and other leading organizations championing the local food movement must stand by and support this new generation of growers. Many growers, new and old, have received low-interest financing for their projects from those institutions that understand the economics behind these efforts. From the implementation of corporate CSA systems to companies offering locally grown food in lunchrooms, meetings and conferences, it is clear that farmers are increasingly gaining larger allies beyond the agricultural industry.

The future is bright for growers who want to grow hydroponically in established greenhouses. Low-cost investment and almost limitless market prospects have encouraged perceptive growers to make a smart step to growing more edibles as a percentage of their overall growing area.

Chapter 16: Common Greenhouse Gardening Mistakes and How to Avoid Them

For the more seasoned gardeners, greenhouse gardening has become a requirement and has been a staple in their gardens. Though some people still make mistakes and we are here to help you avoid them.

Common Mistakes in Greenhouse Gardening and How You Can Prevent Them

Greenhouse gardening is complicated and is not just a simple house where you put your plants in and let it grow. Though the idea of a greenhouse is simple, to absorb the heat from the sun and preserve it for the plants inside, it takes more than just caring for the plants because plants are complex living things too.

Here we mention some common mistakes in greenhouse gardening and some tips on how to prevent them.

1. Forgetting About the Temperature

The optimal temperature of a greenhouse in summer is about 75-85 ° Fahrenheit during the day and 60-76 ° Fahrenheit in the night. During winter, it's 65-70 ° Fahrenheit in the day and 45 ° Fahrenheit at nighttime.

Having a digital thermometer inside the greenhouse will help a lot to track the temperature inside. Having proper ventilation will benefit, as well.

2. Not Providing Shade

A common blunder is by not providing enough shade in a controlled manner while greenhouse gardening. You may have followed the first suggestion, but sometimes the plant gets under stress when it's under direct sunlight for too long.

Anything you can do to fix this is by installing a shade at a specific time of the day or by using a shade cloth to regulate the solar radiation to your greenhouse.

Although you could suggest building a greenhouse near a tree for a natural shade, it can be difficult as there are uncontrollable circumstances that could happen. A tree may let the leaves fall on your greenhouse, the branches may break down and fall on it, or the roots may invade the soil below.

It's avoidable and should be remembered if you create a greenhouse.

3. Poor Ventilation

Good ventilation is essential for greenhouse gardening. But this aspect is still ignored by some people.

It's not just about opening a window, and it's about giving your plants the right flow. Making use of a small fan to keep the air moving will help you a lot.

Air circulation is crucial to the greenhouse, particularly the limited space. Checking temperature and humidity levels is essential to avoid problems with excess soil and leaf moisture.

4. Improper Watering

Too much watering and giving the plants too little water can be a concern and a standard greenhouse gardening error has been made. Drip irrigation is the perfect location for greenhouses.

The trick you can do is to know how much water you need to water your plants in the morning, and then see how many dry plants are there at the end of the day. This will cause the plants to transpire before they get warmer.

Of course, in the warmer season, monitor the plants and regularly water them when needed.

5. Planning Your Fertilizer

Yes, you need to plan and think about fertilization as well. Most beginners should use an all-purpose greenhouse fertilizer, such as 20-20-20.

What is best used for fertilizers, however, are those formulated for greenhouse plants.

6. Location

This is a common mistake and occurs well before someone begins building a greenhouse. The location of a greenhouse is very significant because it essentially covers almost everything mentioned above.

First of all, finding a place to position the greenhouse should be a place where there would be no interference from the sun. The location of trees and buildings should be carefully considered to avoid producing shadows for plants.

Another is that if the location is chosen for the greenhouse to be built, it would be desirable for the house to be placed from east to west to maximize the period of the sun.

Greenhouse gardening is tricky at the beginning, but once you learn everything, it's going to be easy. Know that developing a healthy habit of taking care of your plants is a positive thing.

If you start, it would be helpful to know these mistakes so that you know what to do to avoid them when you start. If you're already into greenhouse gardening and you've made these common mistakes, follow the tips and improve what you've got.

Chapter 17: Tips & Tricks to Get the Most Out of Your Greenhouse Gardening

Enthusiasm is an admirable quality for the beginning greenhouse gardener, but don't let it overwhelm the practical measures you can take to make sure your endeavors are successful.

As with any new enterprise, establishing a strong foundation with careful attention to the smaller details is necessary.

While it will take some patience on your part, this will not only set you up for a positive greenhouse gardening experience, but you will also build an atmosphere for your enthusiasm to continue to flourish!

Keep these seven important tips in mind as you begin your exciting journey to greenhouse garden.

1. Get A Thermometer

Perhaps the most necessary piece of equipment you're buying, your thermometer will tell you when to ventilate, where to shade, when to heat, and what area to cool down.

2. Keep the Greenhouse Clean and Tidy

It may seem like a hassle at the time, but it still takes things back to their proper location. You 're going to be happy you did when you want space to plant pots and when you need a place to grow them. Plus, tons of unused equipment could be used for snails and slugs.

3. Determine Your Plants Space Requirements

Prepare for plant growth by determining how much space your plants would require at the point that you expect them to reach in the greenhouse. Overcrowding can limit plant growth as they struggle for space and light. This stress can make your plants more responsive to disease and insect damage.

4. Start with Seeds Rather Than Seedlings or Cuttings

One way to reduce the risk of problems with insects, mites or diseases is to start planting seeds in your greenhouse — not seedlings or cuttings from another place. It is advised to seed more plants than the amount you want because not all seeds can germinate, and this can help to ensure a strong, healthy plant variety to begin.

5. Start with Easy-To-Grow Plants

Starting your greenhouse growing efforts off with plants that are easy to grow will help boost your confidence. Lettuce, basil and coleus are good vegetative choices. Switch next to tomatoes, seedless cucumber, eggplant, and hot pepper plants when you have more experience. Beginners should avoid starting with bushes and trees, taking more space and taking more time to grow.

6. Take Steps to Keep Pests Away

In addition to using seeds planted or potted inside your greenhouse, mount insect screens on air intakes to reduce pest intake, you should also keep outdoor plants away from your greenhouse. Do some work in your outdoor gardens after you've done your work in the greenhouse, also pest and diseases can enter your greenhouse through your pets so it is advisable to keep them away.

7. Grow Something Greenhouse-Specific

Try your hand to grow something that can only flourish inside a greenhouse, such as tropical flowers or warm-loving vegetables. Not only will this add value to your gardening effort, but the achievement of growing something new will prove to be exhilarating!

Conclusion

Thank you for making it through to the end of *Greenhouse Gardening*, let's hope it was informative and able to provide you with all of the tools you need to achieve your goals whatever they may be.

As you can see, when building a greenhouse or any type of gardening system, there are some issues to keep an eye on or consider. You will cover almost all the problems with the water and nutrient bases, lighting effect and ventilation.

Much of this depends on your grow room, and whether you have access to the outside windows or if it is sealed, and you rely on the uphill lights to the maximum. When you deal with these basic principles, you are able to deal with any system, because the same basic rules are the same.

Once you know the basics, you can easily extend any program, or create a bigger one from scratch. Exploring new ideas and methods is a lot of fun, but these basics will never change.

Do Not Go Yet! One Last Thing to Do!

If you enjoyed this book or found it useful, I'd be very grateful if you'd post a short review. Your support really does make a difference, and I read all the reviews personally so I can get your feedback and make this book even better.

Thanks again for your support!

Hydroponics for Beginners

A Step-by-Step Guide to Quickly Build an Inexpensive Hydroponic Gardening System at Own Home: Discover How to Grow Healthy Vegetables, Fruits & Herbs All Year-Round

Table of Contents

Introduction

Congratulations on purchasing *Hydroponics for Beginners: A Step-by-Step Guide to Quickly Build an Inexpensive Hydroponic Gardening System at Own Home: Discover How to Grow Healthy Vegetables, Fruits & Herbs All-Year-Round*, and thank you for doing so.

I am glad that you have chosen to take this opportunity to discover the great potential and opportunities for Hydroponic cultivation.

I am sure that the information you will find in this book will guide you step by step, and you will learn in a simple and enjoyable way how to approach Hydroponic Gardening.

In this book, we will address every aspect of Hydroponic Gardening, and we will talk about the history, present, and future of this excellent and innovative method of cultivation, we will discover all its advantages and benefits.

In addition, you will learn everything you need to know, from which plants to grow, what equipment you need, what nutrients and lighting you need to install a system at home.

Thanks to this guide, you will have all the useful information you need to know, and you can start growing hydroponics at home with minimal cost. Once completed, you will have everything you need to know to start building and growing hydroponically at home.

Chapter 1: What is Hydroponics Gardening?

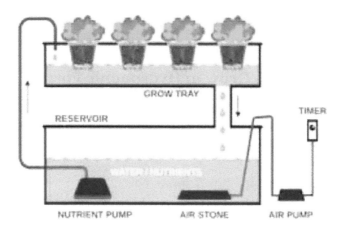

Explanation of Hydroponic Gardening

Plants grow through a process called photosynthesis in which they turn carbon dioxide (a gas in the air) and water into glucose (a type of sugar) and oxygen using sunlight and a chemical within their leaves called chlorophyll. Write that out chemically, and you get this equation:

$6CO_2 + 6H_2O + C_6H_{12}O_6 + 6O_2$.

There's no mention of "soil" anywhere in it — and that's all the proof that you need plants to grow without it. All they need is water and nutrients, all of which are easily obtained from the soil.

Yet if they can get those things elsewhere — say, by standing in a nutrient-rich solution with their roots — they can do without soil altogether. That is the underlying theory behind hydroponics. The term "hydroponics" in principle means plants are growing in water (from two Greek words meaning "water" and "toil"), but because you can grow plants without actually standing them in water, most people interpret the word to mean plants growing without using soil.

Why Do Things Grow in Hydroponic Form?

Hydroponic onions, lettuces, and radishes all grow well. The white surface of these hydroponic containers helps to reflect light evenly on the leaves of the plant, thus growing growth.

Though the benefits of hydroponics have often been questioned, growing without soil seems to have many advantages. Some hydroponic growers have found that when they turn from traditional methods, they get yields several times greater. Since hydroponically grown plants directly dip their roots into nutrient-rich solutions, they get much easier what they need than soil-growing plants, so they require much smaller root systems and can channel more energy into leaf and stem growth. You can plant and grow more plants in the same area with smaller roots and get more yield from the same amount of land (which is especially good news if you grow in a restricted area like a greenhouse or on a balcony or indoor window knowledge). Hydroponic plants are growing more rapidly, too. Many pests are brought in the soil, so usually doing without it will give you a more hygienic growing method with fewer disease problems. Since hydroponics is suitable for indoor cultivation, it can be used to grow plants throughout the year. Automated timer- and computer-controlled systems are making the whole thing a breeze.

This isn't just good news; there are always a few drawbacks. One is the cost of all the required equipment — containers, pumps, lamps, nutrients, etc. The panic aspect of hydroponics is another drawback: there's a certain amount of toil involved. For traditional growing, you can also be very cavalier about how you handle plants, and your plants will always flourish if the environment and other conditions are on your side. Yet hydroponics is more practical, and the plants under your influence are even more so. You have to constantly monitor them to make sure they develop in exactly the conditions they need (though automated systems like lighting timers make it a little easier). Another downside (probably less of a drawback) is that since hydroponic plants have much smaller root systems, they are not always able to be very well supported. Heavy fruiting plants would need pretty elaborate support types.

The History of Hydroponics

You might have seen some kinds of soilless plants grown on movies or books somewhere and viewed it as a story of science fiction.

This method (which is so-called hydroponics) has been, in fact, used for thousands of years.

Babylon's popular Hanging Gardens, in around 600 B.C. Are the earliest Hydroponics documents.

Those gardens were established in Babylonia along the Euphrates River. Since the climate in the area was dry, and seldom saw the rain, people believe the ancient Babylonians used a chain pull device to water the plants in the garden.

Water was drawn from the river in this process and flowed up along the chain system and fell to the steps or landing in the yard.

Many records of Hydroponics in the ancient times were found by the Aztecs in Mexico in the 10th and 11th centuries with floating farms around the island town of Tenochtitlan. And the adventurer, Marco Polo, noted in his writing in the late 13th century that he saw similar floating gardens while journeying to China.

Timeline of Modern Hydroponic Development

It was not until 1600 that scientific experiments on the growth & constituents of plants were reported. With his experiment, the Belgian Jan Van Helmont showed that plants were receiving substances from water. He did not know, however, that plants do require airborne carbon dioxide and oxygen.

In 1699 John Woodward went on to study plant growth using water culture. He found plants grow best in water, which constituted of the most soil. So, he concluded that it was certain that substances in the soil-derived water, which led to the growth of the plant, rather than the water itself.

A number of subsequent experiments were carried out until 1804 when De Saussure suggested that plants be composed of chemical elements absorbed from water, soil, and air.

The French chemist Boussignault went on to test this hypothesis in 1851. He did an experiment to grow plants without soil in an insoluble artificial media that included sand, quartz, and charcoal. He used only nutrients from soil, media, and chemicals. And he found plants require water and get hydrogen from it; the plants' dry matter contains hydrogen plus carbon and oxygen from the air; plants are made of (N) nitrogen and some other plant nutrients.

1860 & 1861 ended a long quest for an important nutrient source for plant production, when two German botanists, Julius von Sachs and Wilhelm Knop, supplied the first basic formula for water-dissolved nutrient solutions in which plants could be grown. This is where "nutriculture" originated. This is also called Water History. Through this process, plant roots were fully submerged in a water solution containing nitrogen (N), phosphorus (P), potassium (K), magnesium (Mg), sulfur (S), and calcium (Ca). They are now known as the macroelements or macronutrients (elements in relatively large quantities required).

Surprisingly enough, however, the process of plants growing in water and nutrient solution was used only as experiments and used solely for plant study in the laboratory.

Only in 1925, when the greenhouse industry emerged, that interest in applying the practice of nutrition was eyed on. Researchers were concerned with the soil culture methods problems with soil structure, fertility, and pests. They worked extensively to apply nutritional benefits to large-scale crop production.

In the early thirties, W.F. Gericke from the University of California at Berkeley has been working with nutriculture for agricultural crop production. He initially named this method aquaculture but dropped it after discovering that this word was used to describe cropping aquatic species.

W.A. A. In 1937, Setchell had recommended Gericke the word "hydroponics." The name goes like this.

The term formed from two Greek words. Hydro ("water") and Ponos ("labor")-literally "hot job." Gericke began to publicize the idea of growing plants in a water solution when he was in the U.C. Berkley. Berkley.

However, he faced public and university skepticism. His colleagues also refused his research using the on-ground greenhouses.

By growing successfully 25-foot tall tomato plants in nutrient-filled solutions, Gericke declared them wrong.

The university still questioned his positive cultivation account and asked two other students to investigate his assertion. The two carried out the study and published their results in a 1938 agricultural journal titled "The Water Culture Process for Growing Plants Without Soil" They acknowledged Hydroponics' application but concluded their work that crops grown with Hydroponics are no better than those grown on quality soils. They ignored many of the benefits of agricultural Hydroponics relative to traditional practice, however. The benefits every hydroponic farmer nowadays knows by heart.

The earliest well-known implementation of Hydroponic plant cultivation was on Wake Island, a soilless island in the Pacific Ocean, in the early 1940s. Pan American Airlines used this island as a refueling stop. The lack of soil has meant that growing with the cultural method is difficult, and the airlifting of fresh vegetables was enormously costly. Hydroponics solved the problems excitingly well and supplied the whole troops on this distant island with fresh vegetables.

Hydroponic farming was also commonly used by the Army during World War II. The U.S. military planted a 22 plantation at Chofu, Japan.

In the 1950s, Hydroponics' soilless approach spread to a number of countries, including England, France, Italy, Spain, Sweden, the USSR, and Israel.

Chapter 2: The Present - Hydroponic Application

With Hydroponics' distinct advantages such as higher growth rate, space saver, water quality, and better pest & disease control, it's no wonder that Hydroponics has been widely implemented around the world.

For any greenhouse grower, it has become an indispensable component.

Virtually all greenhouse farms use some form of Hydroponics to grow their trees & food.

The total area of commercial production of greenhouse vegetables was estimated at 489,214 hectares (1,208,874 acres) according to the International Greenhouse Vegetable Production-Statistics (2017 Edition)

This reports that most countries in the world have constructed vegetable greenhouses, the majority of which are the developed countries, including the United States, Canada, the Netherlands, and Australia. Recently, the majority hydroponic farm (at the time) is being constructed in New Jersey, the United States. They are to carry 2 million pounds per year of fresh, leafy lettuce.

Chapter 3: The Future - Why Use Hydroponics to Agriculture?

Agricultural sectors will face significant challenges in the future as, according to the FAO, food demand is expected to rise by 70 percent in 2050. Given the lack of land, the growing demand for freshwater (farming absorbs 70% of fresh water on earth) and the predicted climate change that can lead to changes in the weather, lighting, as well as the life cycle of plants and animals have to accomplish this.

Hydroponics is, without a doubt, seen as a solution to agriculture's future.

Using no soil, growing fresh vegetables in countries, or any place with little arable land and those whose area size is limited but contains a large population is a valuable cultivation process.

Remote places and tourist sites such as hotels, resorts can hydroponically cultivate their own fresh food rather than importing it from far away regions.

The West Indies and Hawaii are other popular examples. People have used their own vegetable production to serve major tourists. We are going to hope to have more such sites in the coming months.

For water shortage, when desalination technology is in operation, people would be able to draw fresh water from the sea to supply both for the hydroponic garden and for agriculture in general.

Currently, its cost is a big drawback of the soilless planting process. Lights used to grow plants constitute a major part of the cost for a large-scale hydroponic farm. The prices of Hydroponic indoor gardens and those in the northern latitudes with insufficient sunshine are therefore much higher during the year, from late fall to early spring. We believe that growing plants will become much more economically important, with the advent of new technologies in artificial lights.

NASA has found the hydroponic growing system for feeding and nourishing astronauts at the space station and on Mars in the space science industry.

In a world where scientists work day by day to solve food and natural resource problems in a safe and ecological way, Hydroponics still plays a major role in the cope for the future survival of human beings.

Chapter 4: How Hydroponic Gardening Works

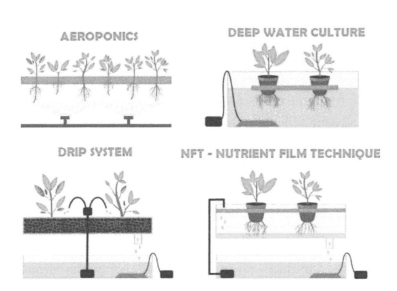

If you've ever put a plant that clips into a glass of water in the hopes of growing roots, you've studied hydroponics. Hydroponics is an agricultural branch where plants are grown without the use of soil. Instead, the nutrients which the plants normally derive from the soil are simply dissolved into water and, depending on the type of hydroponic system used, the roots of the plant are suspended in, flooded or misted with the nutrient solution so that the plant can derive the elements it requires for growth.

The word hydroponics originates from "hydros" in ancient Greek, meaning water, and "ponos," meaning job. Aquaculture or aquaculture may often be incorrectly referred to, but these terms are usually used more specifically for other science branches that have little to do with agriculture.

As our planet's population is growing and arable land available for crop production is decreasing, hydroponics will provide us with a lifeline of sorts and allow us to grow crops in greenhouses or multi-level farm buildings. Where costs of land are already on the high side, crops are already being grown underground, on the rooftops, and in greenhouses using hydroponic methods.

You may want to start a garden so you can grow your own vegetables, but you don't have space in your yard, or you're overwhelmed by insects and pests. This book will give you the information you need to set up a hydroponics garden in your home successfully and provide plant recommendations that will grow readily without a significant investment.

Hydroponics has become a very common solution for urban garden designers due to the lack of soils and space available in towns. Hydroponics is a method for growing food in water without the use of soil, and it can be achieved using vertical spacing, making it a perfect match for balconies in apartments and other confined areas. Otherwise, many ambitious and creative growers have become simply involved in expanding the reach of what they are doing.

In fact, there are many ways of using hydroponics, including wicking beds, floating platforms, flood-and-drain systems, drip systems, and other specific titles that are likely to mean little to those who aren't hackers in gardening. Enough it is to say, hydroponics is a well-established method of farming, something that has been practiced by humans since ancient Egypt. Obviously, we have modern technology that has made things work a little differently, however.

Chapter 5: Hydroponics Plants Basic Concept

Plants can't grow alone on water, so hydroponics is a little more complex than simply cultivating in water pools. Plants also need nutrients, but hydroponic systems don't operate simply on water, but more so on nutrient solutions, water with dissolved minerals, or enriched with rich soils. In pots or small rocks, plants are stabilized so that their roots remain in this solution and are able to feed on the enriched material.

This may seem to be a lot to do when growing plants in the soil have worked well for thousands of years, but under some conditions, all the effort is worthwhile. Hydroponics makes a big difference in arid climates such as Arizona and the Middle East, as it is not dependent on rain and requires less than a quarter of the water soil systems do. It makes the growth of food at home more of a reality in urban areas, where the soil is not available because it can be grown vertically, using less than a quarter of the ground. Greenhouse space can also be used more effectively in northern climes, where cold weather and hours of sunshine are missing.

In short, hydroponics offers an opportunity with less natural resources to reliably generate more food locally, wherever we are.

Different Types of Hydroponics Systems

The good thing about hydroponics is that a lot of different styles of hydroponic systems are available. Any of the best hydroponic systems on the market merge hydroponics of different forms into one hybrid hydroponic system. Hydroponics is interesting in that there are many methods that you can use to get the plants to the nutrient solution.

Deepwater Culture Deepwater Culture (DWC)

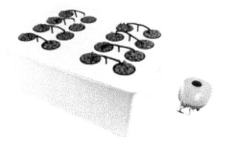

This is also known as the reservoir method, is by far the easiest method for hydroponic plant production. In a hydroponic method of Deepwater Culture, the roots are suspended in a solution of nutrients. The nutrient solution is oxygenated by an aquarium air pump, which keeps the plant roots from drowning. Remember to prevent the penetration of light into your environment, as this can cause algae to expand. This will bring havoc to your system.

The primary advantage of using a Deepwater Culture system is that no drip or spray emitters are required to clog. As hydroponic systems that use organic nutrients are more vulnerable to clogs, this makes DWC an excellent alternative for organic hydroponics.

Nutrient Film Technique

This is a type of hydroponic system in which a continuous nutrient solution flows over the roots of the plants. This form of solution is on a slight tilt to flow with the force of gravity the nutrient solution can.

This type of system works very well, as a plant's roots consume more air oxygen than the nutrient solution itself. Because only the tips of the roots come into contact with the nutrient solution, the plant is in a position to obtain more oxygen, which facilitates a faster growth rate.

Aeroponics

Aeroponics is a hydroponic process by which the roots, when floating in the air, are misted with a nutrient solution. The solution to the exposed roots is obtained by two key methods. The first approach involves misting the roots with a fine spray nozzle. The second approach uses what is known as a fogger bog. If you plan to use a pond fogger, then make sure that you use a Teflon-coated disk, as this will minimize the amount of maintenance required.

You may have learned of the Aero Garden, a marketed aeroponics device. The Aero Garden provides an excellent aeroponic entry point. It is a turn-key system requiring very little setup. To get you going, it comes with great support and supplies too.

Wicking

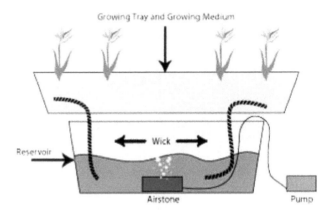

Wicking is among hydroponics' simplest and lowest cost methods. The idea behind this wicking is that you have a substance that is surrounded by a rising medium with one end of the wick substance put in the nutrient solution, such as cotton. The solution then gets wicked to the plant's roots.

This method can be improved by eliminating all of the wick material together and simply using a medium with the ability to wick nutrients to the roots. This works by directly suspending the bottom of the medium in the solution. We suggest using a medium such as perlite or vermiculite. Avoid using mediums like Rockwool, coconut coir, or peat moss because they can consume so much of the nutrient solution that will stifle the plant.

Ebb & Flow

An ebb & hydroponic flow system, also known as a flood and drain system, is a perfect hydroponic system for plant production. This type of system works by flooding the growing area at specified intervals with the nutrient solution. The nutrient solution then flows gradually back into the reservoir. The pump is attached to a timer, so the cycle repeats itself at regular intervals to provide the correct amount of nutrients in your plants.

A hydroponic ebb & flow system is suitable for plants accustomed to dry periods. Some plants thrive when they go through a slightly dry time as it allows the root system to grow larger in search of humidity. As the root system becomes larger, the plant grows faster, as more nutrients can be consumed.

Drip System

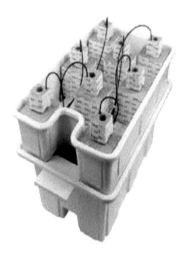

A fairly basic hydroponic drip system. A drip system operates by supplying the hydroponic medium with a slow feed of nutrient solution. We suggest using a slow draining medium, including Rockwool, coconut coir, and probably peat moss.

You could also use a faster draining medium, but a faster dripping emitter would need to be used.

The downside to such a device is that the drippers/emitters are notorious for clogging. We tend not to use drip systems, but if you can avoid the clogs that plague this form of system, it can be an effective method to develop. The reason the device gets clogged is that nutrient particles pile up in the emitter. Systems that use organic nutrients are very likely to run into this type of problem.

Useful Tips for Hydroponic Gardening

I strongly recommend that you adjust the nutrient solution every two to three weeks in your reservoir.

Hold a water temperature between 65 and 75 degrees in your reservoir. The water temperature can be controlled using a water heater or water cooler.

An air pump with an air stone connected by flexible tubing will help to improve circulation and keep oxygenation of your nutrient solution.

If your plant does not look good, either discolored or warped, then the first thing to test and change is the pH. Once you decide that the pH is not the problem, then flush your system with a solution like Clear.

Following the feeding cycle offered by your nutrient manufacturer, we suggest.

Flush, wipe and sterilize all of your environment after an increasing period is over. Drain your tank and remove any debris, then run your whole system with a mixture of non-chlorine bleach and water for about a day. For every gallon of water, use 1/8th of a cup of non-chlorine bleach. Drain your machine then thoroughly and flush it with clean water to avoid any excess bleach.

Why Pick Hydroponics?

Hydroponics is an excellent alternative for cultivators of all forms. It's a great choice because it gives you the ability to monitor the variables carefully, which affects how well your plants grow. A fine-tuned hydroponic system in plant quality and quantity of output will easily exceed a soil-based system.

If you want to plant and grow the biggest, juiciest, most yummy plants that you can imagine, then hydroponics is the right option for you. At first, it may seem overwhelming with all the equipment and effort involved, but once you get the hang of the basics, it will all seem easy enough. Start tiny, keep it easy, and you will never cease to amaze your hydroponic system!

Chapter 6: Hydroponics Vs. Soil Gardening

Saving Space

Compared to conventional soil gardening, Hydroponics saves an enormous amount of energy. In general, the roots of a plant require space to spread through the soil. Today no more! Instead, they get submerged in an oxygenated nutrient solution bath.

Vertical Lettuce Piling-Soil Can't Do That!

Imagine that you got anything in a tiny pill that you had to eat.

You didn't have to search for food or eat three meals a day-you just popped the pill, and your body was dosed with great nutrient supply.

Hydroponics does it for your crop. Hydroponics uses a personalized nutrient solution to provide the plants with precisely balanced fertilizer all the time, rather than using soil as a carrier for the nutrients the plants need.

That is why you have your plants packed closer together, resulting in massive space savings!

Hydroponics Saves Water

Let's remember how the average gardener waters its plants. They normally inject a decent amount of water into their soil every few days to ensure good penetration into the soil so that the roots can suck it up.

Sounds nice, doesn't it?

Well, that is just part of the picture.

Some of the water drips out of their container rim or sinks deeper into the dirt. Many of these evaporate from the surface.

The plant uses only a small percentage of the water. Hydroponics addresses this problem by using what in most forms of systems is called a recirculating nutrient reservoir (Deep Water Culture is one of the most common ones).

This ensures that the roots of a plant can only take up the amount of water they need at any given time and leave the rest for later in the reservoir. To avoid evaporation, the reservoir is sealed so that no water can seep from the bottom.

It makes the same volume of water used for watering a plant in the soil for a day in hydroponics set up for days or weeks at a time to water a plant. By simply converting to a hydroponic system, you can save about 90 percent of the water used in soil gardening.

No Weeding Necessary

One of the most common reasons that I hear when someone asks me why they don't want to garden is: I don't want to weave my hands and knees all of my time!

Simple solution. Convert to Hydroponics. No grass, no grazing. Easy because it is.

Fewer Pests and Diseases

No Soil equals No More of These Bad Boys

Using the same principle, hydroponics dramatically eliminates pests and diseases. Soil is removed from picture and replaced with one of the raising rising hydroponic media. Eliminating soil also removes many of the numerous soil-borne pathogens and pests plaguing conventional gardening.

Double-Headed Time Savings

That's my main explanation for anything. Growing hydroponically not only saves you the time of weeding, pest control, and irrigation, it also accelerates plant production.

If you grow outdoors this ensures you can get in more harvest cycles before the end of your growing season.

You'll also be able to watch plant growth at a faster rate and learn all the different things you might do to make growth even faster.

For example, in hydroponics, you can take a head of lettuce from seedling to harvesting in about a month, compared to two months in soil. Imagine how much more easily you can become a gardening expert with time savings like that!

Gives You Total Control

All of the above factors combine to shape one uber-powerful super factor why hydroponics (and all soilless growing, for that matter) dominates soil gardening absolutely: energy.

You become the master of the environment inside your farm. It is up to you to build the perfect mixture of nutrients, temperature, humidity and that timetable.

It's sort of like that "The Truman Show" series. You're the showrunner and Truman are your plants. You turn and keep the sun out. You decide what you feed your plants, and what they eat. They are entirely responsible for their health. That is a beautiful thing!

You Become a Scientist of The Guerrilla

All the additional control you have over your developing environment makes for a great way to learn how plants can grow. You can adjust the variables and see how they respond to your plants. You get the "environmental recipe" to tailor to whatever plant you create.

Believe me when I say every single plant is special. It is like solving a puzzle to find out what changes you can make to your environment when you develop lettuce vs. basil: you need to make the pieces work together to create a better end product.

That's part of their pleasure! Hydroponics provides you with a small lab to perform any experiments you might think of.

What Would You Expect?

I have given you seven reasons to replace your soil garden with a hydroponic system at the very least. You needn't get big to start. Check out my hydroponic herb guide, if you're stuck for ideas. It will walk you through a simple system that produces year-round fresh herbs! You may also take a look at the best hydroponics books for beginners.

Chapter 7: Benefits and Disadvantages of Hydroponic Gardening

Main Benefits of Hydroponics Gardening

Hydroponics comes with many advantages, the biggest of which is a much higher growth rate in your plants. Your plants can mature up to 25 percent faster with the right setup and yield up to 30 percent more than the same plants grown in soil.

As they won't have to labor as hard to receive nutrients, the plants will grow bigger and quicker. Only a small root system would give the plant just what it needs, so instead of expanding the root system downstairs, the plant would concentrate more on growing upstairs.

All of this is possible by closely controlling your nutrient solution and the pH levels. As the system is enclosed, a hydroponic system will also use less water than soil-based plants, which results in less evaporation. Believe it or not, hydroponics is safer for the environment, as it reduces soil runoff waste and emissions.

Some Disadvantages of Hydroponics Gardening

While there are so many benefits to a hydroponics system, there are also many disadvantages. For most people, the main consideration is that a standard hydroponic device of any size would cost more than its equivalent to the soil. Then again, dirt is not necessarily expensive, and you are getting what you're paying for.

If you're not the most experienced grower, a large-scale hydroponics system will take a lot of time to install. Besides, it will take a lot of time to operate your hydroponics system too. The growing day you will need to track and manage your pH and nutrient levels.

The greatest risk with a hydroponics system is that depending on the size of your device, anything like a pump failure will destroy your plants within hours. They will die fast because the growing medium cannot store water like soil can, so the plants are dependent on freshwater supply.

Hydroponics is not a full-stop solution to food production, despite offering many advantages, especially in particularly problematic areas. Hydroponic systems also have a cost higher initial set-up than soil production. The systems are electrically driven, which makes them vulnerable to power failures, and if that happens, plants will die very quickly in this system. Plus, hydroponic pumps are another tug on the grid if they are not operating on renewable energy. Managing a hydroponic device, fixing pumps, and testing solutions can also be time-consuming, even more so than depending on good soil and rain quality.

Modern hydroponics also has dubious environmental implications for those of us more in-tuned to the traditional ways of doing stuff. Growing in high-quality soils ensures that plants have access not only to the nutrients they need to grow but also to micro-nutrients and trace elements that are essential to maintaining healthy plants and humans. Questions emerged as to whether nutrient solutions would satisfy any of these needs. For hydroponic systems, many traditional crops — potatoes, corn, squash, root veggies, grains — are also not well suited. They're more oriented towards lettuces, tomatoes, and other fairly low, water-hungry bushes.

So, while hydroponics can be very useful, we just shouldn't do away Recovery Drip Systems The recovery drip system is possibly the most common hydroponic system for the small-scale home gardens. This system is designed to improve the efficiency of agricultural irrigation by providing a continuous flow of water and nutrients to the plants and then sending back the unused solution through the network. Many drip designs are set up in such a way that water is fed by gravity through a series of pipes that drain into the pipe below them with a slight incline. Once the nutrient solution enters the bottom of the tank, the water is sent back to the top by a pump, and the cycle continues.

These systems and others like them are really simple to make at home, and if complete DIY isn't in the cards, they are available as kits. It is essential for independent growers not to use PVC tubing, but rather food-grade plastic piping when growing edibles in them, since PVC can leach unwanted chemicals into the solution. It can, however, be a fun, interesting, and successful project to improve food production at home.

How to Begin with Simple Intro Hydroponic Gardening

What is Hydroponics?

Growing plants practically without the use of a conventional dirt medium and using a nutrient-rich water solution. Those media range from fiberglass to sand and fired clay balls to absolutely nothing. Several branches include aeroponics (using air as the medium of growth), aquaponics, etc.

How do I commence?

Okay, you might buy a kit-but it'll cost you. So. Or, to match your needs, you can improvise and build your own set. The cheapest multiplant kit from my local hydroponic supplier is $185, does eight plants, but is not very flexible and very lightweight. Uses the form of ebb and flow. They also sell a single pot bubbler system (bucket) for $50. We will be merging these two systems into a more flexible and much cheaper system.

What Are My Choices?

Several different approaches exist

NFT — stream a thin/flat layer of nutrient solution over the roots) is popular among skilled kits — long with ebb and flow (temporary flooding and drainage of the root system). The most interesting approach is to hang the plants in the middle air and very frequently spray the root system (aka aeroponics). Drip systems are popular as well and have their benefits. MANY methods exist-all of which don't use dirt.

Which Approach Is Used Here?

Bubbler framework is by far the easiest and the cheapest. That is, keep your pots loaded with your medium option just barely above the point of your nutrient solution — then keep the solution aerated well. The air bubbles bursting will keep the typical moist. Recall that simpler and cheaper doesn't mean less effective.

Which Medium is Used Here?

In the past, I used a variety of different mediums. Chopped rockwool, cubes/blocks/slabs of rockwool, fired clay, and a mixture of rockwool and fired clay. This system works best with chopped rockwool (cubed) or fired clay (if starting from seed with this medium extra attention is required).

Cheap?

I'm in college, and I think the cost is high. If you collect parts slowly, this can be a cheap job. And fortunately, the list of pieces isn't long, and they're not uncommon. I think I've spent a total of $30 on new products-but I purchased a few bulk things and splurged a little.

Why Use Water Growing Method?

Foods grown hydroponically not only taste better and are more nutritious, but you can adjust your food's properties, track what's going into your food, and pollute less. In less room, you can grow more too. This is particularly great for those of us with no backyard in which to develop. You can also hold pests away, with the right variety of plants. I want to plant a citronella plant – I don't just like the scent of citronella plants, but their oils keep mosquitoes and other pests away.

Chapter 8: How to Set Up A Hydroponic Garden

Hydroponic gardening means plants growing in a water tank. Many forms of hydroponic gardening systems exist, and some are more complex than others. This style of gardening may be an almost daunting science to get into, but it need not be. Some structures can be put together and managed by just about anybody, with some time and effort. You may choose a simple ebb and flow system, or a wick system, to start a homemade hydroponic garden. Then bring the device together, plant the seeds, and hold the garden.

Step-By-Step Guide to Setting Up Your Hydroponic Garden

Part 1: Putting a Simple Ebb and Flow Device Together

1. Place A Leak-Proof Plate

To start a simple hydroponic ebb and flow method, begin by finding a leak-proof pan. The depth of the pan you are using depends on how many plants you think you are going to grow but should be at least 6 to 8 inches deep to give your plants a growing medium. On the off chance you run out of space in the first pan, you can always use more than one pan. You could use a kitty litter pan for a nice leak-proof pan. You will find a leak-proof pan in a pharmacy, pet shop, or gardening shop to make sure the pan is set in natural daylight, outdoors, or in a greenhouse, or you would need to use a rising light.

2. Line Up Tiny Pots in The Pan

Locate or buy several small pots inside the oven. The seeds are to be planted in those pots. Empty K-cups are perfect for their size, and because they have holes at the bottom of the cups already. Any sort of small pot will do as long as you can poke a few holes at the bottom and sides of it. Depending on the material from which the pot is made, you can poke holes in the pots using a screw. If the pot is made of a tougher material, you'll need to drill some holes.

3. Fill the Rising Medium into The Pots

If the pots inside the pan have lined, fill them with a rising medium. Rising media include hundreds of choices such as gravel, clay pellets, vermiculite, rockwool / stonewool, sand, or cotton and are used to support the root system of the plant as it grows. Ebb and flow systems require good drainage substrates. Gravel or cotton batting can be tried as the medium. When choosing cotton batting, make sure to use an organic brand because cotton is often heavily sprayed with chemicals. Expanded clay pellets work well in ebb and flow systems, too. They have good drainage and can be reused although it is a bit expensive.

4. Flood the Bath

Ebb and flow systems operate on a basic model of floods and drains. For twenty to thirty minutes a time, the plants are flooded periodically every day-the flood process. Then, they empty the tray. People with ebb and flow schemes typically use a submersible pump to do so, operating from a reservoir of nutrients. If you are using a generator, set up your flow and drain system. Most people placed the rising pan in a bigger container above the nutrient reservoir, say, a bucket. The pan and reservoir would then need to be connected to the submersible pump and tubing so that the pump can inject the nutrient solution into the tray. You would also have to mount an overflow pipe to pump the solution back into the tank.

When manually flooding, use at least one cup of water (depending on how many pots there are) and pour over the pan. Make sure the water gets into every bowl. Enable the water to soak in the pots for some time - at least five minutes should be necessary. Drain the excess water into the saucepan by tipping it and allowing the water to drain into a seal.

5. Empty the Pan

The Drain Cycle follows the flood cycle. This is done with a pump, more or less automatically. You can also configure the pump on a timer to operate on. If you do stuff manually, just remove the pots from the saucepan after fifteen minutes of soaking the seeds. Drain into a bucket whatever water is left in the pan, and repeat the cycle many times a day.

Part 2: Constructing a Wick System

1. Find A Tray and Reservoir

Wick systems are possibly the easiest type of hydroponic system to build, as it typically has no moving parts, pumps, or electricity. Via capillary action, the wicking device "wicks" up the nutrient solution from a reservoir to plants in a tray above – in other words, it sorts of sucks up the liquid to the plants like a sponge.

The reservoir and growing tray will be your basic components. Find a leak-proof container that keeps the plants as they expand. That may be bucket, tray, or other container types.

You'll need another leak-proof jar, like a tank, for your reservoir. This container will hold your nutrient solution and should be sufficiently wide to accommodate the rising tray that is normally above it.

2. Pick a Wick

The wick is the distribution mechanism in a wick system – that's what transfers the nutrients from the reservoir below to the above plants, rather than a pump or your own hands as in an ebb and flow system. So, the most critical element possibly is the wick. Your plants do not get the nutrients they need without a good absorbent wick. Common materials that act as wicks include fibrous string, wool, cotton, or rayon cord, tiki torch wicks, wool felt, and strips of old clothing or blankets.

You'll want products tested to see what fits best. Be sure the wick is absorbent, but it does resist rotting. Washing the wick before you use it often also helps in improving wicking performance.

Have enough stuff for wicking on hand too. You will probably need at least two to four wicks unless you have an incredibly limited device.

3. Connecting the Pieces

Since there are no pumps or movable parts, setting up a wick system is relatively simple. Most frequently, people place the rising tray directly above the tank and attach the two to the wicks. In fact, it's best to get those parts as close as you can – the shorter the wick, the more water it will bring to the expanding medium of your plants. Next, you'll need to punch holes at the top of your reservoir and at the bottom of your tray. Then, bring the containers in place and thread through your wicks.

Seek to uniformly distribute the wicks over the bottom of the rising tray.

Finally, apply your rising medium to the tray's bottom so it can cover the wicks. Wick systems require an absorbent medium such as vermiculite, coconut coir, or perlite. Always make sure to wash out the medium about every two weeks with fresh water, because this will reduce the possibility of building up to harmful levels of nutrients and salts.

Part 3: Plant the Seeds

1. Place the Seed in Each Pot

When you have set up the device, you are able to plant the seeds. Your preference is what sort of seed you choose to plant. You will grow several flowers, herbs (such as basil and thyme), and vegetables (such as spinach, lettuce, and kale). Put one seed into each saucepan. Enable the seeds to soak in the water that you poured into the pots for about fifteen minutes. In a hydroponic system, even the beans grow well. The seeds normally germinate within 8-10 days.

2. Select Your Plants as A Nutrient

To grow and survive, plants need a full spectrum of nutrients. When the seeds start growing plants, you'll need to select a nutrient to make sure everyone is receiving what they need. For a healthy hydroponic greenhouse, that's key.

Plants need 16 elements to grow at appropriate concentrations. Getting too much or very little of any nutrient can lead to poor crop yield. That said, it's best to look for a commercial hydroponic solution that provides a complete nutrient profile. Hydroponic nutrient solutions come in two basic forms: driven and liquid. As a novice, you may want to continue with something a little more error-proof in the liquid solution. Some are more costly but don't require mixing. 3 Pull or transplant the plant. You should halt until the plants are fully grown to harvest them. The time for the plants to grow depends on what you have planted. Plants growing in gravel or other hydroponic media aren't easy to transplant, so many growers wait until they are fully mature and harvest them all at once. Wait until the bed is dry to cut the plant and shake off any particles that might still be attached.

Part 4: Maintaining Your Garden

1. Get A Grow Lamp

You may need a grow light during the winter, or if your plants are not put outdoors in a garden or greenhouse. A growth light mimics natural daylight. They can be bought at gardening shops or online. Some plants need lighter than others, so study the amount of light needed for growth for each plant you're planting. You can control the amount of light your plants get with a simple timer that regulates the on / off setting of your grow light. An optical timer will work just fine. A digital timer is not required.

2. Test the pH Level

You will check the pH level of your garden periodically. This can be easily achieved by picking up Nitra zine paper, which is available in many drugstores. To use, simply dip one of the strips into the nutrient solution you are using and compare it with the chart that comes with the paper. By adding soluble potash or phosphoric acid to the nutrient solution, depending on the test result, you can maintain a pH level of between six and seven.

3. Using Pest Insecticide Soap

Even the hydroponic gardens are vulnerable to pests. You may use an insecticide soap or pyrethrin based spray to get rid of pests. You can purchase any of these pesticides from most gardening shops or online. Make sure to follow the instructions on the pesticide type label you want to use.

4. If You Find A Disease, Sterilize the Bed

Other signs of plant disease include bleeding, blighting, rotting, and tumors. If you experience any of those signs, use a diluted copper spray or sterilize your yard. Remove the pots, temporarily transfer them to another container to sterilize your greenhouse, and spray the original container with a diluted bleach solution. Enable the bleach to sit for 24 hours, then empty the bottle. Then, wash several times thoroughly with water.

Chapter 9: Ways to Set Up A Hydroponics System

Many people have either already started with a hydroponic garden or are looking to start one. While there are large quantities of knowledge available, when it comes to operating or constructing a hydroponic system, many sources do not appeal to new growers.

There are several types of systems that users can use, but these fundamentals remain the same irrespective of which system is being used.

Here we'll look at each section in turn and see how it reflects on your program, or how it can influence your decision about which program you'd like to create. We'll offer an overview of the most common types of hydroponic systems people want to operate before digging into the core fundamentals. It can be purchased or leased, and no matter how the device is designed, the end result is the same.

Hydroponic Systems Forms

For all these systems, we will discuss what they contain without specifying nutrients or pH kits because they will be common for all systems.

Kratky System

Difficulty: Beginner

It is one of the easiest hydroponics methods available and does not require electricity. All these requirements are a dark container with a lid, net pots, and you're through newspapers. These are suitable for growing spinach, lettuce, tomatoes, and many other leafy vegetable types.

The holes in the lid will be where your net pots will be set. When they're filled with the growing media, the young roots protrude into the solution from the bottom of the net bowl. As nutrients are absorbed, the level decreases, and the roots expand. The gap in the top of the container at this stage provides oxygen to the roots.

Deep Water Culture

Difficulty: Beginner

A DWC system is quite similar but on a bigger scale to the Kratky system. This system needs only an aerator pump and can work for a greater number of plants. The water continuously flows into this device to provide oxygen to the plant's roots.

Flood and Drain (Ebb & Flow)

Difficulty: Intermediate

It is one of the most common hydroponic systems in use, and it can, for a good reason, produce some of the best results out of all the systems while being easy to maintain. Here the plants are cultivated within pots in any suitable growing environment. These are then lined up in a rising field, a little higher around the sides than the containers.

The reservoir of nutrients will be a separate unit, large enough to contain enough nutrient solution to flooding the rising bed. There's an overflow pipe in the grow bed that will stop water from growing above the pots' height. Next, you will need a water pump to fill the grow bed at intervals, after the timer is done (at intervals of 15 or 30 minutes) the water flows back into the reservoir via the pump.

Plants get enough space, and oxygen absorbs until the bed develops. The addition of air stones will provide the roots with more oxygen, which leads to healthier plants.

NFT (Nutrient Film Technique)

Difficulty: Advanced

Such processes are all in one solution. In most cases, they contain wide 4-inch PVC pipes that have holes cut into them that hold your net pots, they are again filled with your rising media, and the roots are protruding into the pipe's bottom.

The reservoir will carry the pump and air stones which will flow continuously down the pipes and return to the reservoir. Roots are still exposed to oxygen in the soil so they won't get overwatered. The vulnerable spot is where a pump failure happens. Plants will fail quickly because their growing medium won't retain enough moisture to sustain the plants for prolonged periods.

Drip Systems

Difficulty: Intermediate

Each plant will be fed individually inside this system. The plants are fed from above, while all other systems are fed from below, which differs in this type of system. There is also a separate tank containing both water pump and aerator stone, and tiny diameter tubes from the pump feed the pots over a timed cycle. The nutrient then drips back to the reservoir from the bottom of the pot or tub.

It is very similar in nature to a flood and drain network but without the element of flooding.

There is another method called aeroponics, which differs from all of these because it sprays a fine mist of nutrients onto the roots. They may be much more complex to set up, and might not produce as many plants in your growing room. However, many of these systems are purchased rather than installed.

A Look at Hydroponic Basics

We'll now look at the hydroponic basics. This will include knowledge seen beforehand, which will be universal across all systems.

There may be more problems with one method than another, but there'll be the same methods to solve them. For example, if you have waterlogged roots, the solution would be the same regardless of the method. In hydroponic systems, all that differs is the process of nutrient distribution and ease of service.

Chapter 10: Light, Air, Water & Hydroponic Nutrients

The Importance of Light

When it comes to growing in this kind of system, germination needs very different rules, so they will have their own portion, however, once transplanted plants do light, which is the most important ingredient for plant growth.

If you've your system in an outdoor place, or an indoor grow space, without light, and enough of it, your plants will not thrive, doesn't matter. Light is intended to promote photosynthesis where carbon dioxide and water are converted into plant food and oxygen. So, the faster and more bountiful your plants can grow, the more light they get.

There is one important note, and that is if you grow indoors under artificial light, you will produce better results than if you grow plants in a sunny window. It is because you monitor the amount of light in the day as opposed to the variable light.

The sunlight varies throughout the year, and even though your plants are warm enough, their growth will continue to change throughout the seasons.

The one thing you must ensure when using grow lights is that plants obtain the full spectrum of light. It has to be all the way from the bluer end of the continuum to the other where red light abounds.

There are countless types and strengths which growers choose when using grow lights. From fluorescent, Lead, high-pressure sodium or metal halide lamps, you can choose from. Many of which come with particular benefits and downsides.

What's important is that regardless of which bulb you're using, all of them are governed by the following three factors:

CCT (Correlated Color Temperature)

It is measured using a Kelvin scale and will apply to the light source temperature. That does not automatically mean that one bulb is hotter to the touch than the other. It is determined by the temperature of the sun.

A good example is that 6000 Kelvin is known as cold, and a light source that has a 2700 CCT (Kelvin) will be called warm.

Wattage

A use of rising lights watt is how much energy it can use when turned on. This also has a connection to how bright the electric running through the product would be.

A typical at-home light bulb uses around 75 watts. Still, there are lights of 250, 600 up to a 1,000-watt bulb being used in some hydroponic grow rooms. They will use the same energy at this stage as you are A / C or some other bigger electrical unit.

The disadvantage of most of these HID lamps is that they need to change periodically: 24 hours a day – change after six months 18 hours a day – change after nine months 12 hours a day – change annually.

Lumens

That is how much light that your light bulbs can emit per square foot. The manufacturer controls this calculation and not power consumption (wattage). High lumens are the equivalent of high-quality lamps. Always settle for a lamp delivering less than 2,500 lumens. For each square foot of your increasing area, you can also use the measurement of 20-50 watts.

With all this, it is important that you provide at least 14-16 hours of good light to your plants every day.

The Key Role of Air

Found under climate is a few variables, which would be humidity, temperature, and carbon dioxide. All of these are critical for any hydroponic system and can have a serious impact if they are not in the right place, or if they do not reach the appropriate standards.

Each breath that we exhale releases carbon dioxide, and to some degree, there are levels in the air we breathe in. Both rates are regulated in outdoor gardens without too much interference, but that can be very different for an indoor garden.

Instead of only providing airflow, air ventilation is important because the CO_2 levels may only reach about 400 ppm (parts per million). The required rates will be up to approximately 1,500 ppm, and it will need fresh air to reach your growing room without unnaturally supplementing this.

When using grow tents, you can see how some growers manage this, this growing the area, so CO_2 ppm increases. Such tents will have an intake fan, and the other side will have a fan that blows in the fresh air.

We come up to the ambient temperature after this. Depending on the plants, this will alter. Nearly all plants are unable to live outside their ideal growing temperatures for too long, however. This can be either up or down, but farmers shouldn't presume that it doesn't get too cold simply because they're indoors.

It is another environment influenced by an intrusion of fresh air from a ventilation system, as well as any through lights in operation. Careful monitoring and adjustment of the ambient air temperature are necessary.

The next area to be influenced by pollution is moisture. Some plants are highly selective on humidity levels. It is suitable for mold or algae when humidity is too high, and this is particularly true in your reservoir. Low humidity, on the other end, as it struggles to absorb moisture, may stress plants.

Water and Hydroponic Nutrients

The different requirements of each device.

This may be attributed to the farmers or plants being grown. Many growers already prefer mixed nutrients, so it is enough for other people to add these in varying quantities simply.

However, there is an underlying aspect of this that is important to the productive growth that growers should be aware of. It is easy to buy a bottle of nutrients, but understanding what they are doing is very different.

It is useful to see how commercial growers use their powdered compounds to produce their desired results, understanding how these elements work.

NPK is what all growers know to follow to optimize their plant production. These come in various solutions, like 20-10-5, for example. That means there is 20% Nitrogen, 10% Phosphorous, and 5% Potassium.

When plants reach later growth, lower amounts of N (Nitrogen) should be obtained, which is why some larger-scale commercial growers mix their own, they are in a much better position to manage each compound as required.

One thing to remember here is that it correlates back to you develop room temperature. If the temperature is below 80F when your plants are in their vegetative growth phases, you may need higher amounts on N (Nitrogen). If your rising room has a higher temperature than this 80F, then you do not need to change the N levels.

There are plenty of NPK ratios you can see, but not all of them are going to be perfect for your garden. Here is a guideline that can be used as a starting point: N (Nitrogen) – 200 – 400, P (Phosphorous) – 200 – 600, and K (Potassium) 200 – 600 Often plants growing have magnesium deficiencies. An easy cure for this is the application of Epsom salts to your bath. Regardless of how you get your nutrients, and that can be dry or liquid, so never over-feeding your plants is the one basic thing. It is much easier to underfeed slightly and strive to popular rather than add.

If we look at nutrients and water, we have to remember the temperature, and if your mix is too dry, this can lead to bacterial growth. Certain areas that require attention and are important for good plant growth are EC (Electric Conductivity Factor) CF (Conductivity Factor) TDS (Total Dissolved Salts) pH Levels If you look at your TDS levels, they are optimal when they fall within the range of 500 – 1000ppm in your solution. The ppm plants can take up nutrients more easily when there is a decrease. In younger plants or seedlings, this is true, and once they reach their stage of vegetative growth, these rates can be increased to 800 – 900 ppm. This would be equated with a better approach to the nutrient.

When plants achieve their period of maturity or flowering, these levels of TDS can be further increased to 1000ppm – 1100ppm before being decreased in the final stage.

Whatever levels you use, these levels of TDS should be calculated on a continuous basis along with the pH levels. EC meters measure the solution's conductivity and will calculate the TDS, which is the dissolved salt amount.

We all know that pH levels need to be regularly assessed. These function in tandem with your solution's strength as they can shift when you add nutrients or add fresh water to your tank. Ignoring these can have a drastic impact on your plants or leaving them too long before testing.

One essential aspect is never to mix pH UP or pH Down into your mixture of nutrients. A chemical reaction can occur with the chemical components. This is the chance; you should use only one dropper for every chemical. Once it is applied to your tank, water should be tested for modification, so you can adjust without disturbing your plants with a sudden rise or fall in pH.

To ensure that your nutrients do not become too solid, you can complement (top off) your reservoir with half the nutrient strength of what you began with first. This could be done every other day, and you can use plain water (pH checked) to top up the days in between.

Changing Nutrients and Flushing Reservoir

This would be one of the most important things for their system that a grower needs to do. Still, specific growers use two ways to determine when to do this. One is using a time-based approach, and the other is using the volume-based approach. That is where the thresholds above for TDS come into play.

Here are the two main ways of doing this:

Time-Based

This can be between 10-14 days. It demanded that all of your old mixtures of nutrients be disposed of and refilled with a new solution. Larger systems can be extended to a month before the approach shifts. It is the preferred way to gauge when to adjust from the two.

Volume-Based Flushing

The time to flush the device using the above TDS and EC readings is when you've used half the starting number. This way, when nutrients are consumed, many people top off their tanks. This involves constant control, as soon as the top of number hits half of the original starting volume, then it is time for the machine to flush out.

Growing Mediums and Water Quality

A rising medium contributes little in the way of nutrients to a hydroponic system. The only role there is to provide help for your plants. Some of these growing mediums are appropriate, and a grower may use more than one form on occasion.

To use, many growth media have to be thoroughly soaked. It is the case when germinating for Rockwool but also important when used in a complete method. Usage of the best growing medium is important for the plants you grow and the type of device your usage. A strong example of this is the coco coir. The fibers can start blocking the pump in systems that use water pumps, but in DWC or NFT, they have less chance of being easily moved through the system.

It needs to be sterile when using water for your hydroponic system because it does not contain any harmful elements. That's why farmers use water that is pure or reverse osmosis. This also ensures when it comes to adding their nutrients, they have a clean slate.

This means you have hard water when you see sediment buildups around your faucets. If that's the case, then it means that you have too many minerals. When you use this form of water, and it has a TDS of more than 200 ppm, you will need to adapt to accommodate this change, because it will impact your TDS with added nutrients. Perhaps that is the justification for using clean water.

We have shown that maintaining the nutrient mix at a given temperature is crucial. It should be noted, however, that optimum root growth should occur when your mix is between 70-75F, but if you have any root diseases, they should grow faster under these conditions.

If you hold a temperature below 68F, this will go a long way to help avoid root rot in your plants.

Additionally, the temperatures in both DWC and other 'Bubbler' systems are more difficult to control as there is normally no external reservoir and low amounts of water. Another argument that is very often ignored is that aquarium pumps can run hot, so they add heat to your mix of nutrients.

Chapter 11: What to Do and What to Avoid for Your Hydroponic System

It's known that certain plant varieties do not grow very well in hydroponic systems. Most root crops are not as productive as leafy vegetables, due to how growing media works in comparison with soil.

There are, as we said, several rules that need to be adhered to, as well as a handful of stuff that you can do and stuff that you shouldn't. You can see these here:

What to Do

Light exposure to your reservoir should be limited to when you finish off your solution. To avoid light exposure, all pipes that lead to rising beds should be dark, and any possible way light will pass through your cover should be capped.

Wash the supply lines and drain frequently. This can be every couple of days, depending on the type of program you are using.

Over time, pumps lose capacity, test these to ensure they work as they should. In particular the flood and drain systems.

Clean the filters when they start clogging. Coco coir can be causing blockages here.

If you are using reusable through paper, wash it before re-use, and let it dry.

Perform comprehensive medium and pot sterilization and any other components during growing cycles.

Sanitize the components and the machine every day.

Do test the pH levels all the time. And if you are using a litmus strip for a quick test, you can easily see if any changes occur.

What to Avoid

Never use lemon juice. Never use liquid bleach in your system. Never use iron while running some UV, which can lead to the formation of chelates. Some UV should be turned off before cycling of the device.

Don't let your rising media stay wet; it will result in root rot. Time the cycles, so before the next feeding cycle, the rising media is almost empty.

Never treat rough water with a water softener. These add salts harmful to plants, to the water.

Do not cause your nutrient mix to rise above 75F or go below 50F. To compensate for this, you would need an aquarium heater or chiller. Air stones may be used to help refrigerate as long as fresh air is used.

Don't put lights for growth too close to your plants. If you have no choice, point a fan that oscillates to where the hottest areas will be.

Choosing Plants

One of the most popular questions for hydroponic beginner gardeners is: "What can I grow? "The simple answer is that you can grow any plant hydroponically, provided the correct setup and nutrient balance. In order to select which plants will be better suited to your home system, you should consider the following factors: what type of system you have or want to create, how much room you have, how much experience you have, and why you chose hydroponics.

What Sort of Device Do You Have?

Hydroponic systems are either Solution, or Liquid Culture and Medium, or Aggregate Culture. The plants grow exactly in the nutrient-filled solution in a Solution system, such as Aeroponics or Nutrient Film Technique. With plants that are fast-growing and shallow-rooted such as lettuce, spinach, radishes, and herbs, this form of setup works best.

Medium systems like Wick Systems or Ebb & Flow systems use a through medium like gravel, sand, or Hydroton. Since the medium provides good support for heavy plants, these setups work well for vegetables and herbs with deep roots such as comfrey, chicory, and beets, or those which are heavier and need support such as beans, tomatoes, squash, and cucumbers.

If you do not yet have a hydroponic setup, considering what kind of plants you would like to develop will help inform what kind of system you want. If you are in love with fresh herbs and salads but have only a limited amount of room to devote to your garden, you can do your best with a limited D.I.Y. Wick Device. If you're more experienced and are searching for more exotic options or a way to improve your hydroponic garden, you may want to consider going for a more high-tech system like Aeroponics, for example.

How Much Is the Available Space?

When picking which plants to grow in your hydro-garden, space is a major factor to consider. If you have only a limited area to devote to your garden, you will do your utmost to avoid tomatoes, melons, and other big plants. While you could theoretically grow these in a small system, on your plants, you will never get the same quality of fruits or vegetables as those with ample room to grow. In small systems, leafy greens and herbs are the simplest and most rewarding options. These are plants that grow fast, that can be harvested on an ongoing basis, and that does not need much room to expand or produce fruit.

If you have a wide room, such as a greenhouse, garage, or patio, you can switch to a more advanced system and grow those voluminous plants that need trellises and deep root support. Large gardens are ideal for experimentation — you could even try growing your favorite fruits and vegetables in several different varieties.

What Is Your Level of Experience?

Your level of gardening experience and your particular reasons for getting into hydroponics will also influence your choice of growing plants you wish to grow. When you are an absolute beginner, it would be best to stick to fast-growing, simple plants so that without being frustrated, you can get the full gain from your experience. Even though hydroponic gardens are simple and easy to maintain once you know the basics, if you start with an overly complex method, it is easy to become discouraged.

When you're experienced in hydro-gardening and want to venture into more exotic or complex plants, the only true limit is yourself. Some gardeners also go so far as to plant whole fruit or nut trees hydroponically. The sky is the limit if you have space and the will to explore. Quince, cotton, big melons, pumpkins, sunflowers, and shrubs such as honeysuckle or blackberries are a few suggestions for more experienced gardeners.

Chapter 12: Equipment Needed for Hydroponic Garden & Which to Choose Between Different Types of Hydroponic Systems?

Necessary Tools

- Bucket
- Garden Hose
- pH Test Kit (Including pH Buffer Solutions)

Required Materials

- Plant Clips
- 50-Gallon Nutrient Tank
- PVC Pipe
- Plastic Tubing
- Extended Clay Pebbles
- Water Planting Cups and Trellis Made of PVC Pipe
- Fertilizer
- Twine
- Plants

What Different Hydroponic System Types Would You Use?

Ebb & Flow Systems

Ebb & Flow systems use a tank of water that is kept separate from where the tanks of herbs are. A pump pulls the water into the herb containers so that they can collect the water and nutrients they need, and then the water flows out from the containers into the main tank, which is then properly balanced, purified and supplied with nutrients.

Deep Water Systems

Considering hydroponics, these systems are planned with the beginner in mind. The water in this device stays oxygenated and circulated with the use of a small pump, which is ideal for the growth of herbs.

Aeroponic Systems

The aeroponic systems are used for optimum oxygen access to herbal roots. These systems use a method of misting spray that allows the roots to obtain the water and nutrients required but leaves them exposed to the essential oxygen in the air.

Drip Systems Drip systems pump water and nutrients into the herbs' containers in smaller amounts for a sterile approach; a timer keeps the drip system on track and operating at regular intervals.

Chapter 13: Growing Plants and Herbs with Hydroponics

If you grow herbs for culinary or medicinal purposes, it doesn't matter; hydroponics is a perfect way to grow them. There are several reasons to do so, and the first is that they are rising faster. You can also add to this that they come with more flavor and fragrance than counterparts produced in soil do. Research also shows that hydroponic herbs contain more aromatic oils up to 40 percent.

Not only this, but growers can grow a variety of herbs that they would otherwise be struggling to grow in their own field.

Hydroponics makes these herbs simple to grow:

- Basil
- Chamomile
- Rosemary
- Oregano
- Cilantro
- Anise
- Dill

Like all plants, herbs care about temperature, light, and water. If you swing too low or extremely high for any of those herbs in either direction, they will end up dying. Growing herbs using hydroponics helps you keep yielding herbs, whatever the weather or season. Hydroponic development takes up only less space and reduces water use.

Although all herbs can be simple to grow in a hydroponic system, here are the top eight herbs to cultivate. We're going to go through the basics and benefits of every.

Basil

Basil is a common option for hydroponics since this herb is suitable to hold on to the aroma and flavor when used fresh. Those attributes are lacking on dried basil. And seeing restaurants and greenhouses using a hydroponics device for their basil herbs is not unusual.

There are 150 different basil species in total, but the most common ones are Sweet Basil, Genovese Basil, Thai Sweet Basil, Purple Basil, Lemon Basil, Lettuce Basil and Spicy Basil.

Basil can be planted in two ways, by germinating the seeds, or by planting cuttings that form their roots within a week. Basil is a warm-weather herb, so holding temperature between 70-80 Fahrenheit is best. Blocks of rockwool are the most common media used in hydroponics with increasing basil. While you can use peat moss, coconut coir, perlite, and vermiculite, these will require sterilization prior to use.

Pythium is a threat to Basil seedlings, you should remember. What exactly is Pythium? Pythium is a fungus that attacks many herbaceous crops and spreads disease. The best way to prevent Pythium or other damping-off pathogens is to make sure that the media surface is not too humid.

When you get to harvest basil, the top 1/3 to 2/3's of the upper foliage can be cut. The plant will keep growing that back so that you can cut it again. Basil will regrow up to 2-3 times before removing the plant altogether, and starting fresh is recommended.

Just cut the quantity of basil you need; this saves the stress of trying to keep it in good shape. When you pick basil, basil's shelf life is only a few days, so it might be best to keep it growing on the plant before it is essential.

Chamomile

If you're a big tea fan, you may want to learn you can grow your own Chamomile with hydroponics indoors. Chamomile has many impressive antioxidant properties, which have been shown to reduce the risk of diseases such as heart disease and cancer. Often, they help combat insomnia and poor digestive problems.

Most would use a floating seed tray to help the chamomile seeds germinate. You will want to get rid of the weakest ones after the seedlings grow to around 2 inches, and there's just one good seedling per cell in the tray. It can take up to 1-2 weeks for chamomile seed to germinate. Chamomile is advised to receive up to 16 hours of light every day.

Chamomile has wide compatibility, as it relates to pH ranges. This can range from 5.6 to 7.5 in anywhere. Ideally, you'll probably want to reach 6.5 in the middle for optimal results to rise. You will be able to harvest your chamomile flowers after about eight weeks.

The flowers can be picked by cutting off up to 3 inches of stem and then drying them in a sunny area on a cloth. Through not picking all the flowers, you will make replanting much easier, which helps them to re-seed themselves. For protection, store your Chamomile in an airtight container in a dark, cool place. You can read more here to see more of the benefits to Chamomile's health.

Rosemary

This is a Mediterranean evergreen herb, with leaves that look like needles. The herb may flower in white, pink, purple, or sometimes blue. Rosemary can be used as an aid to a wide range of issues such as stomach problems, Heartburn lack of appetite Cough Headache High blood pressure Low blood pressure Toothache Insect repellent and more rosemary growing hydroponically compared to other herbs will prove much slower. You should expect a harvest time of up to 12 weeks, and the seed yields are often extremely low. They still prove much more effective to expand hydroponically.

Such plants are susceptible to infections with the fungus, powdery mildew, and mites. The most suitable for this herb is an NFT hydroponic system, and they should be subjected to temperatures ranging from 70 degrees Fahrenheit to 85 degrees Fahrenheit max.

Here are some fast tips to hydroponically grow the rosemary.

Keep the pH range from 5.5-7.0 Moisture levels will remain normal.

Expose the herb to 11 hrs. of daylight as a minimum. You can harvest 2-3 times per sowing, and this can be achieved during the year.

Oregano

Oregano is of the mint family, and for thousands of years, they have used this herb for cooking and medicinal needs. Oregano was used by the ancient Greeks for treating G.I. problems, menstrual cramps, urinary tract infections, skin conditions, and dandruff. Many times, they have studied Oregano for its antimicrobial activity that wards off pathogenic Listeria.

Hydroponic Oregano can grow well in pH levels from 6.0 to 9.0, and the level should fall between 6.0 and 8.0 for optimum performance. Rockwool cubes are widely used to germinate the seeds that can take anywhere from 1 to 3 weeks. Some other rising media are the Rapid Rooters, Oasis Root Cubes, or Grodan Stonewool.

Oregano is a slow grower, and after a transplant, it can take up to 8 weeks before the first harvest. Oregano likes full sun when you grow outside, and when you grow under the lights, the lighting won't be any different. T5 tubes are suitable for providing the right light, and they should be around 2 to 4 inches from the tops of the plant to prevent drying or burning the leaves.

Cilantro

From seed to harvest, when grown hydroponically, you're looking at some 50-55 days for cilantro. This choice of the herb is very low maintenance and needs no trimming. They can be harvested in part or in full.

If you are a food lover, you already know what healthy cilantro is. Toppings, garnishes, salsas, this is what you call it. Although some people don't like the taste, why? A lot of people perceive the cilantro taste differently. Some qualify it as a fresh and cool taste, while others consider soap-like to their tastes. Here's a scientific explanation of why this is. A few tips to hydroponically grow cilantro: Keep the pH level between 6.5 and 6.7.

Temperatures of up to 75 degrees Fahrenheit will stay anywhere between 40 degrees Fahrenheit. Nevertheless, for temperatures in the 60s, there are higher germination speeds.

Watch out for spots of powdery mildew and bacterial leaf, popular to cilantro. Such spots cause high humidity levels and exposure to too much humidity.

It does need plenty of water, but it doesn't have to be overwatered. Also, it is recommended that oscillating air recreate a sturdier outdoor environment.

Anise

This uncommonly heard of the herb has a taste of licorice. Often, it is also called aniseed. While Anise can fight off many common problems such as digestion, gas, cramps, and more, other herbs also help.

Although the taste of the liquorice sort may make it unpopular with others, it is resourceful in savoring bread, sausages, cookies, and cakes. Anise seedlings are very delicate and difficult to move, so it is best to allow the seeds to germinate and grow without moving them in their respective containers. You'll find the seeds can germinate for up to 2 weeks.

You're going to want to maintain a pH level of about 5.5 to 6.5. Meeting at 6.0 in the middle is the most ideal for development. The seedlings benefit the most from having an oscillating fan stirring the wind gently for a few hours each day.

The best way to harvest Anise is to cut the plant as required, and place it in a protected area free of direct sunlight to dry out. It can hang them upside down, too. They are harvested entirely as soon as the heads start becoming brown—store away from heat and light in an airtight jar. Anise usually has a shelf-life of up to 1 year.

Dill

Dill is an annual growing herb in the family of celery. It's most frequently seen grown in Eurasia, where it is used to flavor milk. For your recipes, you can use the fresh dill or dried dill. The stems aren't used when using fresh dill. Growing dill is quite simple hydroponically, and thrives in this type of growing environment.

Culinary uses for dill include: Soups Salads Dips Casseroles Pickles Medicinal uses include: Relieving stomach bloat and gas Headaches Cramping Catnip If you have a cat, you may want to grow this herb hydroponically mainly for their enjoyment and, of course, to provide yourself with some mild entertainment. Catnip is not only used for cats, somewhat contrary to common perceptions and the name itself. Catnip has been known since the early 1700s for its capacity to alleviate cramps and indigestion when used in herbal teas.

Here are some tips for growing catnip hydroponically indoors: By using leaf-tip cuttings or seeds, you can easily propagate catnip.

Provide up to 5 hours of daylight.

Using good drainage to provide a constant volume of water. Catnip can be susceptible to root rot, so try avoiding an area that is too muddy.

Look out for the growth of molds, which can occur from too much misting.

Eliminate any insect infestations, including aphids, mealybugs, scale, and whitefly.

Don't let that cat come close to your system!

Hydroponic Growth Tips

Growing the seeds on a piece of rockwool and press them in.

Keep the Rockwool moist with nutrients and water waiting for the seeds to germinate. Germination can take 7-10 days but may occur earlier.

You can then put the Rockwool directly into your hydroponic system after germination. Maintain a pH range from 5.5 to 7.5.

Enable sufficient space to grow, and note that sometimes dill will actually grow as high as three feet.

Harvest only by cutting the leafy leaves and removing the stems when brown and ripe seeds appear.

Is It Easier to Hydroponically Cultivate Herbs?

As always, when it comes to efficiently growing plants and crops, hydroponic systems emerge as top contenders do. Herbs can benefit the most from the capacity of the watering system to obtain a continuous supply of nutrients and oxygen. In a hydroponic climate, on average, herbs grow about 25 percent to 50 percent faster than an outdoor soil climate.

Besides, some herbs are better off being young. Hydroponic systems offer their customers the opportunity to cultivate fresh herbs for restaurants, supermarkets, and commercial farmers, which allows for greater flavor and cost-efficiency.

Find these advantages of hydroponically growing your herbs:

You don't need any soil. Although some that love the naturist appeal of dirtying your hands by gardening out the sunlight, the truth is, some of us prefer not to have to go that direction. Hydroponic growth really requires only some water and clear mediums.

You can get bigger yields and quicker growths. As previously mentioned, in a hydroponic system, you would see 25 to 50 percent faster growth than you would as an outdoor crop. This quick development means you'll be able to yield more in less time.

Less upkeep. Most hydroponic systems are running on autopilot, so you can only test the pH balance and update the nutrient solution regularly.

Herbs are most commonly prey to insects and pests. The ownership of an indoor hydroponic device would greatly remove such risks.

You'll have more space to save. On average, hydroponic systems only use up to 10 percent of the water used by outdoor soil plants. The water is filtered and continuously reused.

You can monitor the surrounding. Is your region vulnerable to floods, storms, or even temperatures that are frigid? With an indoor hydroponic device, which will still be in a tightly regulated and secure environment, you don't have to worry about this.

You don't need to use herbicides and insecticides, ensuring you can keep your plants 100% safe and free from harmful chemicals.

Using hydroponic gardening, you'll save a huge amount of space. Systems can be tailored and even vertically installed.

Some say hydroponic gardening helps with stress relief. There's always something to put within your home a part of your outdoor world. Having another living breathing element close, you can have positive effects on mental health.

Getting into it is an all-around fun sport, what's better than the happiness you get from realizing you've grown a plant from start to finish, nurturing it every step of the way? If you have a natural green thumb or not, hydroponics is straightforward, for beginners too!

Chapter 14: Growing Fruits and Vegetables with Hydroponics

Growing Hydroponic Fruits and Vegetables Nearly any plant or vegetable will grow hydroponically. The concerns you need to ask yourself are: Why do you want it to grow? What makes you own a hydroponic garden? How big is that unit? So how many components does it have?

In case you intend to make use of your soilless garden for a hobby or to pass the time, go ahead and enjoy it. Plant something that suits you; do not hesitate to try. In mathematics two decades ago, the level of knowledge about hydroponics today is around the same as that. We need to know about the niche so that you can help others too. And the pros of research and experiment continuously. The only criteria I believe is always to have fun. Try it all out.

My advice would always be to stick mostly to vegetables for those who are very serious about the crops they would like to harvest. This is primarily the salad vegetables that industrial farmers have altered through hybridization until there is a considerable lack of these initial nutrient values and flavors. A few examples include plastic lettuce, swampy tomatoes, soggy radishes, and hollow celery.

You will, of course, be limited to how much space, time, and money you need to devote to the whole notion. It is here that practical considerations will come into action. In the sixteen by the twenty-four-inch tub, for example, six tomato blossoms containing six pounds of tomatoes each. The container reflects a much more effective usage of space compared to 16 corn stalks.

The following guidelines are given for helping home growers of hydroponic vegetables. (It also included a few fruits that could be grown hydroponically.) Knowledge of nutrient requirements is useful for those individuals who grow by themselves. There are a few things common to bear in mind. When growing a few veggies species in one tank and adequately using an industrial nutritional supplement, care should be taken not to disturb the balance. Note also that when seeding or transplanting into your soilless garden, the whole area can be used to grow and the constraint on how far away your seeds are depending on the physiological environment around the plant we need to grow. A pea blossom climbing up a string, for example, requires much less air gap compared to a bushy tomato plant.

Common Fruits and Vegetables

Beans

Beans may grow indoors or outdoors, in winter or summer. Bush beans grow indoors in winter. Pole beans grow outdoors during summers. The polar varieties could be tied and cultivated vertically.

Often, they are planted very close together (about six inches). As its title suggests, bush beans appear to take up just more of a spot. However, beans need less oxygen than other plants and need greater quantities of potassium, phosphorus, and sulfur. Lima's don't produce a crop as big as they do, plus they take even longer to grow.

Cabbage

I grew cabbage without making it head over. Pick on the leaves for dinner as you would for leaf lettuce, and let it continue to rise. Plant six centimeters. Cabbage needs cool weather and high levels of both nitrogen, carbon, and phosphorus.

Carrots

Because of the depth of the rising medium, gourmet carrots are better to produce than the ordinary varieties. Plant roughly ½ inches apart. Potassium is essential and phosphorus.

Cauliflower

For some cause I have had bad luck with cauliflower. It's particularly resistant to fluctuations in temperature. If you grow cauliflower with other plants, it is best to cultivate it with moderately cool requirements along with plants. Plant some 8 inches apart. Larger quantities of nitrogen, iron, and phosphorus are required.

Celery

This is truly an excellent vegetable salad to rise. On the other side, celery does well and even hate high temperatures. Plant about 4 inches apart, and have your salad with the young stalks and leaves. It is best for around two months and a thin pencil. It would only be good for sauces and stews by the time it is four months old. Do not uproot a whole plant at one time; just cut off a number of stalks. Larger amounts of chlorine and sodium are typically all essential.

Chard

This is really an outstanding crop that can be harvested just like head lettuce. Keep the outside leaves removed for your dinner. Plant them four inches apart and keep temperatures high. Chard is a nice cooked veggie, like spinach.

Beets

The Beets root vegetables are best grown with a relatively limited soaking in vermiculite. To reduce the buildup of algae, only a slight coating of haydite or gravel should be used. Most beet varieties are great. They excel at cool temperatures. Plant just a few centimeters. Grow smaller beets for increased tenderness, and the majority of these.

Broccoli

A number of experts agree that this is a great crop. We will use transplants, spaced seven inches apart. Broccoli takes advantage of the cool weather (60 ° F, 16 ° C). Enormous quantities of nitrogen, carbon, and phosphorus are essential.

However, Corn is a potential crop; it's not dry, but due to the small produce. Plant some six inches apart midget grain.

Cucumbers

This is a preferred commercial crop along with lettuce and tomatoes. You don't want to cross-pollinate, plant the English, or the seedless kind in case you do. These develop well indoors or in greenhouses, but if you grow them outdoors and insects do the pollinating, some oddly shaped cukes can end up having. They prefer hot weather and direct sunlight and are vulnerable to molding at times.

Leek

You can obtain a superior harvest by adding more potassium
and nitrogen, and extra phosphorus.

Eggplant

But this may not be a real favorite crop. Eggplants are slow germinators and like the warm weather. In case you pull a few of the blossoms away, they will grow bigger, allowing just a few nice fruits per plant. Excessive quantities of nitrogen, potassium, and phosphorus are needed, but if possible lower the nitrogen after the fruit has already grown.

Lettuce

Boston and New York are very common varieties of heading, but leaf lettuce yields far greater harvests. When you're growing head lettuce, get rid of the outer salad leaves without even having to head for it so you can maximize your harvest. Grand Rapids is wonderful foliage lettuce along with Salad Bowl, but note for Caesar Salad the Romaine (Romagna). An abundance of lettuce can be obtained in six weeks or less. But, the first two weeks need to be treated with caution. Lettuce will be bolted with inefficient high or low temperatures (small leaves will expand to a long, stringy stem) Varieties not bolting, such as Black-Seeded

Simpson, Endive, Escarole, and Batavia. For this particular harvest, it would be prudent to cut back a bit to your nutrient source. Lettuce enjoys mild temperatures (50-70 ° F, 10-21 ° C) with high levels of nitrogen. Place them about four inches apart, near the edges of your planter, so their heads are hanging over them.

Melons

Melons growing methods are very similar to those for cucumbers. They want to be, day and night, in warm weather. High humidity induces mold, and thus keeps it very well ventilated. Honey Dew is always a magnificent cantaloupe, of course, if you'd rather try watermelon, take an early variety such as baby Sugar. Just note Cross-Pollinate. Tie up the vines, and provide plenty of lighting when growing indoors.

Onions

Common are Spring Onions or Green Bunching. They should be sown quite strongly, at a distance of one-half inch. Needs a greater potassium and nitrogen content.

Peas

Both types excel in hydroponics but use their sweet and flavorful edible pods to try Snow Peas. Act with a whole lot of plants to get some great harvests. Link them, or allow a trellis to mature. Plant back three inches and hold to cool temperatures.

Peppers

Every pepper was good for growing: Green Bell, Yellow Banana, or Chile. Grow them together or on their own. Peppers want hot weather even more. Put them apart for six inches and watch for damping off. Peppers are much harder to grow indoors than outdoors because they want high light levels that aren't reliably reached by indoor light. Our experience is that peppers and tomatoes don't like each other; when grown alongside peppers, the tomatoes avoid growing.

Radishes

Other types are all appropriate, but like beets, it is much easier to grow them in vermiculite and to plant approximately one and a half centimeters away. Hold the vermiculite as moist as possible around halfway. Radishes bolt easily, so make sure they've got enough warm and cool temperatures. Water can be used only for the very first few weeks when radishes are grown alone. Generally speaking, radishes are grown in the worst part of a garden, but in hydroponics, they have the best of all, and if you're not very careful, you'll get a lot of tops before the root can grow.

Spinach

Spinach is a fast-growing crop—plant two to three centimeters. Cold temperatures and a lot of nitrogen are absolutely necessary.

Squash and Zucchini

These plants are grown just like cucumbers, just note how much room a zucchini plant occupies and then plant between eight and nine inches apart. Between seven or six sets of leaves pinch the plant off to conserve the energy closer into the root and to guarantee seed.

Strawberries

These are very good for delivery, but not too cost-effective if you don't intercrop. Seek to get a range as self-pollinating as Ozark Beauty. Place them apart for eight inches and sit back for a long time. Strawberry plants require two to three years to mature, much like asparagus.

Tomatoes

While the tomato is actually a fruit, it is usually counted among the vegetables. It is among the hydroponic plants that are stronger & more satisfying. You should seed bush or tomato patio indoors, so your plants are likely to stay far below your lights. Often you can extend staking tomatoes; the bush variety remains easier to work in hydroponics, particularly if the vines have not yet finished growing whenever you are ready to bring them in at the end of the summer.

Crop tomatoes to pick outdoors early and late. Plant the seeds under lighting in February or even in March for your initial tomato harvest and transfer them outdoors in April or May.

Other Vegetables

A variety of other vegetables that you may want to grow

Basically, you can grow something outdoors, no matter how far you can stretch your own vineyards. You will only be able to expand anything you can light up indoors, so you will also be much better off sticking into varieties of bush, stunt, or patio that will still be under your lights. Other plant types may need to be pruned as they expand, or their vines grow too fast. We recommend you just focus on these plants inside, such as lettuce, tomatoes, other salads, fruits, and herbs-all the items that give nourishment at a time when it's most needed and is the supermarket's priciest.

Outside it makes sense to take full advantage of the available hydroponic growing area. That can be achieved by intercropping and outcropping—the rapid growth of plants among slower growing species. A rapidly growing crop, like radishes or leaf lettuce, would have appeared, and space and time needed for a slower-growing crop has been harvested by now.

Outcropping involves having the crop spread out, up, down and sideways, from the field. The model shown below gives you a clear idea of how to accumulate growth and produce far more than the accessible growing area seems to allow.

There are a few reminders in place. When you want to grow root vegetables, such as carrots and radishes, you'll only find two things to note. Next, irrigate them after only planting them with simple water for the first week or two, before they have proven themselves to be very small, stocky plants. To the water, just apply nutrients. Second, due to the medium's relatively shallow depth, you must not grow such a thing with a root significantly longer than three inches. This is not an issue with round radishes, just the herecle variety, and there are also small, barrel-shaped varieties of carrots over the seed shelf.

Companion Planting

Plants don't make a sound, so you would think that their environment was peace and harmony. Not so-there are other enemies and friends amongst the plants. Some crops defend one another from infestations of insects, but others give their friends some shade. Also, if they're friends, some actually like each other and grow stronger. In hydroponics, you're likely to want to learn at least two plants that grow together happily.

Garden Bouquets and Houseplants

Anything that blossoms inside a dirt garden or flower pot would do much better in a hydroponic planter, from asters to zinnias, summer or winter. The same applies to house plants. They are Tropic children and live mostly in a state of permanent hibernation throughout our latitudes. Both seeds and transplants perform extremely well in hydroponics, plus it's awesome to see them grow significantly in the same way as they would in the tropics in a comfortable climate. Houseplants do not use as much water as vegetables or flowers, but your plants can never be overwatered due to the outstanding aeration properties of a hydroponic medium. This is certainly one of the potted house plants' normal causes of death. When sending flowers, you're probably much better off using varieties that grow up to 9 or even 12-inch because they grow twice as big as hydroponics. Flowers that emerge from land gardens in excess of 1-2 inches would be the hydroponically grown too unfavorable. Lots of laws relating to vegetables would also apply to plants in homes.

Get rid of the plant from the tub and then wash off the soil gently using cold tap water from the roots. The cold water helps to anesthetize against jolt the vine. Add a suitable growing medium for the container, and the plant will possibly sit at the same depth as before.

Using clear drinking water for 10 to 14 days in a bottle. It pushes the key outlet in its brand-new world to sell and expand in its search for the dietary supplement. Start the nutrient alternative at 10 to 14 days, or earlier in case the leaves turn light green or yellow. Label your calendar and bring it to use for a single calendar month once you start using nutrient response. For a month, turn to plain water, then flush the plant with barely lukewarm water to ensure it remains good and fresh for your plant life. This removes salt and mineral buildup, which, even from the medium, appears as a white crystalline formation. Enable emptying and adding fresh water. Should not use water from the boiler because it's too plump. Keep on using water the next month for alternate nutrient remedies together.

If you transplant a flowering plant, it will probably shed most of its buds and blossoms, but in all probability, it will still be alive.

Chapter 15: Hydroponics Gardening Issues and Mistakes to Avoid

If you're just commencing with your hydroponic garden, you want to take it slowly and easily. One mistake will ruin all the progress you made on growing up. Rather, take the time to consider what you expect of your plants and the conditions they require.

Common Issues and Mistakes

There may be many issues in a hydroponic garden, but here are the five most popular mistakes a hydroponic gardener may make:

MISTAKE 1 – Overlooking pH levels

The most important metric for your hydroponic system is its pH level. Thanks to a nutrient solution, the plants survive almost entirely for the most part. If the solution is too alkaline or too acidic, the plants will suffer deficiencies in nutrients or just die.

Get a top-notch pH meter and watch the rates at least once a day. If it is moving in one direction or another, take immediate action to restore it to the balance that your plants like.

An off-kilter pH level in a hydroponic system is one of the most frequent causes of plant die-off. Monitoring pH levels is extremely important because all of your plants live in the same nutrient solution-if your pH is bad for one plant; all of your plants might suffer!

MISTAKE 2 – Purchasing Cheap, Incorrect, or Not Enough Lighting

You can make or break your hydroponic garden by investing in the right light! If you buy too little, it will damage your plants. If you buy your plants the wrong type of bulb, they won't grow. If you want to purchase the cheapest bulbs, they may not be able to work.

Lighting is one of the major investments that you are going to make as a hydroponic grower, so find the best for your crop! This means you should investigate what kind of light your plants would need because different bulbs are sending out different types of energy.

If they are set next to a window, don't expect your plants to thrive either. That light is often not strong enough to fuel the vigorous growth that a hydroponic plant expects of you.

MISTAKE 3 – Use False Plant Food

Buying a sack of fertilizer from your local garden center for use in your hydroponic system can be tempting. It's all about nutrient distribution, after all, right?

Not so! Not so. It is not likely that conventional fertilizer can dilute fully in your system. It may also clog drains and pipes. Instead, invest in hydroponic systems-designed fertilizers. Hydroponic fertilizer, which is available as liquids or granules, fulfills the rising needs you need in a soilless or soil-light garden by supplying additional nutrients that your plants can otherwise lack.

MISTAKE 4 – Not Focusing on Sanitation

Don't let your hydroponic garden area become a garbage bin. Your sanitation practices can have a big effect on plant safety and all of your hydroponic system.

Any basic cleaning needs that you should address: keeping the floors clean and dry Sterilizing and cleaning system equipment Sterilizing and cleaning devices Sterilizing and cleaning containers Disposing of plant waste Without proper sanitation, you can spread plant disease or provide hiding places and sustenance for pests.

MISTAKE 5 – Opting Not to Know

Traditional hydroponic systems have been around since the early 20th century, so a lot of knowledge, so the advice is made available during that period. College courses are taught on this subject. There are hundreds of books in-store. You will also find plenty of hydroponic growing knowledge with daily updates in our Articles section on the Safer ® Brand here.

In short, do not go for it by yourself! Read up, and make a hydroponic garden plan—converse and swap thoughts with other hydroponic gardeners. The more you learn, the happier you will be by the time you are able to harvest before setting up your garden.

Tricks and Trips to Make the Most Out of Your Hydroponic Garden

Here are the very best hydroponic gardening tips I've gained from several years of personal (often the hard way) practice. I've seen a lot of people try hydroponic gardening once or twice and struggle to try, never again. In general, the explanation falls into one of three groups.

Lack of knowledge — you don't know how things should be or what you need to do without discipline — you know how things should be, and you know what needs to be done, but you don't take the time or make an effort — you need more first-hand experience, or maybe you don't have the requisite hydroponic gardening equipment or supplies. It took two years of making mistakes and doing stuff the hard way before I changed my attitude and took those lessons to heart, even after reading advice like this myself. As a result, I had the most productive garden I had ever seen. Follow the tips below to shave the learning curve years off, and skip straight to excellent results!

Chapter 16: Best Hydroponic Gardening Tips

Know what equipment you need and why

Know the nutritional requirements of your plants

Know the light/photoperiod requirements of your plants

Use a qualified three-part hydroponic nutrient product

Do not use additional nutrient additives the first time

Have a detailed schedule before you start.

Having a Plan

Having a plan means understanding the nutritional and photoperiod requirements of your plants, and having the supplies and equipment required to fulfill those needs. Actually, providing a detailed week-by-week feeding plan, along with nutrient strengths and nutrient adjustments, will also be very helpful. these feeding instructions also come with the General Hydroponics and B.C. nutrient starter kits. Ingredients.

Food / Nutrients

Know your plants' nutritional requirements before beginning. Know how high the nutrients are expected to be in your plant's life every week, and know what the nutrients will consist of each week. At first, many plants need more Nitrogen, then move to need more Phosphorus to grow fruit or flowers. Get a TDS or E.C. meter to keep track of the strength of your nutrient solution, and change it as your plants grow.

Don't risk mixing up your own vegetable food. Start with a qualified commodity of hydroponic nutrients, instead. These are typically detailed (and simple to use) three-part systems. My preference with Technaflora is B.C. Nutrients. When your hydroponic gardening system is up and running and produces outstanding results, then if you like, you can try to combine your own special plant food. At least then you'll know exactly what the problem is when things work out!

For the use of nutrient additives, the same applies. Don't seek to boost your performance by adding a bunch of extra items (at least not at first) to your nutrient reservoir. This is one of the hydroponic gardening tips I end up explaining to everyone that I personally assist! Start by feeding only the basic nutrients in three parts until your hydroponic gardening system works smoothly and yields excellent results. You can then try adding vitamin B1, liquid seaweed, or Silica (or all three) if you prefer.

Finally, your nutrient reservoir has to be tested and kept regular. You need to start anew with fresh water and fresh nutrients after two weeks of using the same nutrients. The most effective way to do so is to provide two tanks of nutrients, one with a nutrient solution for your hydroponic growing method and one with plain water for changing the next nutrient. I can't emphasize the value of the hydroponic gardening tip! The second reservoir helps the water to de-Chlorinate and hit room temperature, all of which protect your roots. See my Nutrients Hydroponic page as well.

Root Health

If the roots get damaged, they cannot take nutrients for plant feeding. Any damage below the ground as dead leaves and sick plants can result in damage over land. Protect your roots by properly maintaining your nutrient solution, using two hydroponic nutrient reservoirs (one with plain water for your next change in the nutrient), and minimizing the amount of light that comes into contact with your nutrient. This will prevent algae, which will prevent fungus gnats, preventing most problems with root damage.

Adequate Lighting

When it comes to lighting up an indoor garden, very few shortcuts. You'll need at least 40 watts / sq. Ft., but 60 watts / sq. Ft. It must have been great. Either high-pressure sodium lights or metal halide lights are the most common option and will do a very fine job. I suggest either a 600-watt light or a 1000-watt light for various purposes. Try my Light selector app to help you choose a lamp. This will be one of your biggest expenses-plan to pay on average between $400 and $600 for a good machine (light+reflector+ballast).

A few hydroponic gardening tips on fluorescent lights: normal fluorescent lights do not provide ample useful light for healthy growth and are only suitable for vegetative clones, seedlings, or very young plants (spinach, lettuce, cooking herbs). If you want to go in your grow room with fluorescent lights, T5 lights (aka Tek lights) really are the only way to get there. Although T5 lights produce less heat than HID lights, they yield per watt only around half as much. It is also important to keep the tops of your plants within a few inches of the light, which sometimes turns into a real pain in the ass.

Temperature Management

One of the best tips on hydroponic gardening is temperature management! Plant development STOPS rapidly when the temperature rises above 85 degrees (unless you are actively pumping CO_2). HID grow lights drive a lot of heat out, which makes temperature regulation a major problem for the indoor garden. Placing the ballast outside the grow room for your light will help (only if you have an old magnetic coil ballast, and not a nice new digital ballast lol), but that's not nearly sufficient. An utter must are centrifugal fans or squirrel cage fans (see exhaust fan setup). Fans alone are sometimes not enough, in my experience. What's really needed is a cool/cold air source.

I have come up with just two hydroponic gardening tips after years of organic and hydroponic gardening to solve this problem ... Expect indoor gardening if the outside temperature is 55 * F or less. In this way, when you expel the hot air, you can draw cool, dry air into your garden (from an outside source). The only other choice is air conditioner pumping!

Manipulation of the Photoperiod

Several crops need shorter periods of daylight to cause blooming / fruiting. Here are two hydroponic gardening tips for success: First, the lights have to be switched off and back on at exactly the same time every day (use a remote timer!).

Second, during the dark time, the plants should be held in full, complete, UN-interrupted obscurity. Either use a complete black-out room dedicated to the garden or use a black-out growing tent. Plants can be very prone to this, so don't start skirting around that! Check out my floral forcing page for more details.

The Right Equipment / Tools

Don't start your garden unless you've covered all your bases right from the start. You will need a fully dark area, a high-powered fan, sufficient ventilation, a hydroponic gardening system, hydroponic nutrients, an oscillating ventilator, a TDS meter (or E.C. meter), a pH test kit, and probably an air conditioner. Minimum speed. I promise a thermometer and a digital timer will come very handy too!

Conclusion

Thank you for making it through to the end of Hydroponics for Beginners, let's hope it was informative and able to provide you with all of the tools you need to achieve your goals whatever they may be.

As you can see, when setting up a hydroponic device, there are certain issues you need to keep an eye on or consider. You'll cover nearly all with the basics of your water and nutrients, the lighting effect, and ventilation.

Much of this depends on your growing room, and whether you have access to external windows or it is sealed off, and you rely on rising lights to the maximum. When you are in charge of these basic principles, you are in a position to take on any hydroponic system, since the same basic rules are the same.

Once you know the basics, you can extend any program easily, or create a larger one from scratch. The exploration of new ideas and methods is a lot of fun, but these basics will never change.

Do Not Go Yet! One Last Thing to Do!

If you enjoyed this book or found it useful, I'd be very grateful if you'd post a short review. Your support really does make a difference, and I read all the reviews personally so I can get your feedback and make this book even better.

Thanks again for your support!